Gender and Gentrification

This book explores how gentrification often reinforces traditional gender roles and spatial constructions during the process of reshaping the labour, housing, commercial and policy landscapes of the city. It focuses in particular on the impact of gentrification on women and racialized men, exploring how gentrification increases the cost of living, serves to narrow housing choices, makes social reproduction more expensive, and limits the scope of the democratic process. This has resulted in the displacement of many of the phenomena once considered to be the emancipatory hallmarks of gentrification, such as gayborhoods. The book explores the role of gentrification in the larger social processes through which gender is continually reconstituted. In so doing, it makes clear that the negative effects of gentrification are far more wide-ranging than popularly understood, and makes recommendations for renewed activism and policy that places gender at its core.

This is valuable reading for students, researchers, and activists interested in social and economic geography, city planning, gender studies, urban studies, sociology, and cultural studies.

Winifred Curran is an Associate Professor at the Department of Geography, DePaul University, Chicago, IL, USA.

Routledge Critical Studies in Urbanism and the City

This series offers a forum for cutting-edge and original research that explores different aspects of the city. Titles within this series critically engage with, question and challenge contemporary theory and concepts to extend current debates and pave the way for new critical perspectives on the city. This series explores a range of social, political, economic, cultural and spatial concepts, offering innovative and vibrant contributions, international perspectives and interdisciplinary engagements with the city from across the social sciences and humanities.

For a full list of titles in this series, please visit www.routledge.com/Routledge-Critical-Studies-in-Urbanism-and-the-City/book-series/RSCUC

Urban Subversion and the Creative City
Oliver Mould

Mega-Event Mobilities
A Critical Analysis
Edited by Noel B. Salazar, Christiane Timmerman, Johan Wets,
Luana Gama Gato and Sarah Van den Broucke

Art and the City
Worlding the Discussion through a Critical Artscape
Edited by Julie Ren and Jason Luger

Gentrification as a Global Strategy
Neil Smith and Beyond
Edited by Abel Albet and Núria Benach

Gender and Gentrification
Winifred Curran

Gender and Gentrification

Winifred Curran

Routledge
Taylor & Francis Group

LONDON AND NEW YORK

First published 2018
by Routledge

2 Park Square, Milton Park, Abingdon, Oxfordshire OX14 4RN
52 Vanderbilt Avenue, New York, NY 10017

Routledge is an imprint of the Taylor & Francis Group, an informa business

First issued in paperback 2019

British Library Cataloguing in Publication Data
A catalogue record for this book is available from the British Library

Library of Congress Cataloging in Publication Data
Names: Curran, Winifred, author.
Title: Gender and gentrification / Winifred Curran.
Description: Abingdon, Oxon ; New York, NY : Routledge, 2017. | Series:
Routledge critical studies in urbanism and the city | Includes bibliographical
references and index.
Identifiers: LCCN 2017008256 | ISBN 9781138195844 (hardback) |
ISBN 9781315638157 (ebook)
Subjects: LCSH: Gentrification--Social aspects. | Urbanization--Social
aspects. | Women--Social conditions. | Sociology, Urban.
Classification: LCC HT170 .C87 2017 | DDC 307.76--dc23
LC record available at https://lccn.loc.gov/2017008256

ISBN: 978-1-138-19584-4 (hbk)
ISBN: 978-0-367-36202-7 (pbk)

Typeset in Times New Roman
by Taylor & Francis Books

Contents

Figures

Acknowledgements

This book is the result of twenty years of research on gentrification. It would not be possible without the academic colleagues, anti-gentrification activists, friends, and family who helped to shape my thinking and support me along the way. Many thanks to my writing buddies, Alex Papadopoulos, Colleen Doody, Heidi Nast, and Andrea Craft, for their support and erudition; Alison Mountz who has been unfailingly supportive and encouraging; Euan Hague, who gave me the space to get this done; and Trina Hamilton who is both collaborator and friend. Chicago is chock full of urbanists who have inspired me along the way, especially Rachel Weber, Janet Smith, Brenda Parker, and Alec Brownlow, with special thanks to Michelle Boyd, who both read my book proposal with great care and provided great wisdom on the process of writing. Leslie Kern and Heather McLean have been instrumental to my thinking on these issues. Victoria Romero, Alejandra Ibañez, and Nelson Soza invigorated my faith in the possibilities of resistance. Thanks to Mike Jersha for allowing me to include his map in the book. My thanks to my editor at Routledge, Faye Leerink, who helped guide me through the process, along with Priscilla Corbett and Richard Skipper, and to the two reviewers whose invaluable comments helped strengthen the final product. Research was partly funded by a University Research Council grant from DePaul University. Much of this book was written at one of Chicago Public Library's great institutions, Sulzer Regional. Support your local library.

Mothering heightened my interest in and experience of gender in urban space. Thank you to Phil and Siobhan, who have accompanied me on many an urban expedition and even sometimes let me get work done. My infinite gratitude to my sister and best friend, Mary Curran, and to my parents, Judith and Thomas Curran, who taught me how it's done.

1 Introduction

Despite an extensive academic literature, and even more extensive, if often inaccurate, press coverage, there is much about gentrification that remains hidden. As Schulman (2012: 28) comments, "Since the mirror of gentrification is representation in popular culture, increasingly only the gentrified get their stories told in mass ways. They look in the mirror and think it's a window." Gender is one of these often hidden aspects. For much of the over 50 years since Ruth Glass coined the term, the debate has focused on whether the cause of gentrification is primarily the result of the production of new urban spaces to satisfy capital's need for accumulation, or rather, the result of an increased demand for urban space from the new middle class in an era of deindustrialization (see the back and forth between Smith (1979; 1987) and Ley (1986, 1987) and the degree to which displacement actually occurs when a neighborhood is gentrified (see Atkinson 2000, Freeman and Braconi 2002, Freeman 2006, Newman and Wyly 2006; Slater 2006, Wyly et al. 2010; Vigdor 2002).

Around the 50th anniversary of gentrification as a term, there have been any number of reevaluations of what it means and where research on gentrification should be headed. Lees (2016) has called for increased attention to race and ethnicity, especially in the Global South, and indeed for more attention to gentrification in the Global South more generally (Lees et al. 2015). Brown-Saracino (2016) calls for more attention to what Lees (2003) termed super-gentrification, and calls on researchers to explore the connection between gentrification and the increasing concentration of poverty (though, of course, Smith (1996) did this work by explicitly placing gentrification within the process of uneven development). Much recent work has also called into question the very importance of gentrification as a process within larger urban trends and the continuation of suburbanization in the U.S. and elsewhere. Zukin (2016) terms it "unimportant" in this context. Studies which focus on quantitative measures to locate gentrification, such as those by NYU's Furman Center (Dastrup et al. 2015) and the Philadelphia Federal Reserve (Ding et al. 2015) find it limited to fewer neighborhoods than most qualitative studies would suggest. City planners in Chicago have told the anti-gentrification activists in the Pilsen neighborhood (with whom I have been working for 12 years) that gentrification is not a problem.

This book is my contribution to the debate on where we have been and where we should go. Following Smith (2002), I see gentrification as a global urban process. I am far less concerned with how we define or quantify it than I am with contesting gentrification, preventing and ameliorating its negative consequences. Gentrification is producing a highly unequal city in which a redistribution of resources to certain up and coming neighborhoods results in the increasing ossification of poverty in disinvested neighborhoods. In this focus on effects, I argue that the gendered nature of these effects has been too little examined since the flurry of work around gender and gentrification in the 1980s and early 1990s (e.g. Markusen 1981; Rose 1984; Bondi 1991a; Bondi 1999). Notable exceptions to this trend are Kern 2007, 2010a, 2010b, 2013; van den Berg 2012, 2013. When gender is addressed in the research, it tends to be with women as gentrifiers rather than as the gentrified. The willful invisibility to gender, both in research and in citation practices, in the greater body of gentrification research reinforces the patriarchal practices that allow for the continuation of those gendered effects and helps to foreclose potential avenues for resistance.

While gender is not necessarily central to the definition of gentrification, it has been central to gentrification's effects in a patriarchal, heteronormative system. For the purpose of this book, I understand gentrification as the influx of upper income uses into previously working class neighborhoods resulting in the displacement of those working-class users. I like Schulman's (2012: 14) definition of gentrification as a "concrete replacement process. Physically, it is an urban phenomena: the removal of communities of diverse classes, ethnicities, races, sexualities, languages, and points of view from the central neighborhoods of cities, and their replacement by more homogenized groups." This definition recognizes the multiplicity of changes that this class transition accomplishes. Gentrification represents not just a change in the housing market. It forestalls the possibility of other ways of being for the populations in urban neighborhoods before these are "rediscovered."

Displacement is central to the experience of gentrification, though as more recent research understands, this displacement need not always be physical (Anguelovski 2015; Davidson 2008, 2009; Shaw and Hagemans 2015). The displacements of gentrification are central to the theme of this book. Class is gendered, raced, aged, and abled. In "reclaiming" parts of the city that had previously experienced disinvestment and displacing the populations who lived in those areas, gentrification exposes the fundamental inequalities of modern society and urban planning. Gender is foremost among these, though it is rarely recognized as such in the gentrification literature. The goal of this book is to daylight the many and varied effects of gentrification on the social construction of gender, for "while gender identities may be fluid, diverse and ultimately impossible to generalise, particular modes of gender power may be named and traced with some precision at a relatively general level" (Brown 2006 quoted in Jupp 2014: 1311). I contend that these effects are overwhelmingly negative, reinforcing gendered divisions of labor, privatizing public life, and atomizing groups within the city. These gendered effects are so common they have become banal, uncommented upon in most gentrification research. My goal in putting gender at the forefront is to raise a feminist consciousness and critique that can create solidarity and strength, building resistance that recognizes common violences and the structures that produce them (Parker 2016b: 2).

It wasn't supposed to be this way. Gender was central to early debates about the causes and visible effects of gentrification, and some theorized that gentrification could be emancipatory. Markusen (1981) went so far as to argue that gentrification was a result of the breakdown in the patriarchal household. Gentrification, so theorists argued, had the potential to allow for a more flexible and visible role for groups such as women and gays in the urban environment, a reworking of gendered assumptions, and thus a fundamental change in urban form (e.g. Caulfield 1989; Rose 1984). Evidence of this sort of demographic shift was partly supported by work by Smith (1996), Rose (1989) and Mills (1988) who found the following key features in gentrified neighborhoods in New York, Toronto, and Vancouver, respectively: a female population increasing faster than the male population, a high proportion of young and single women, a high proportion of professional women, high levels of academic credentials, a high proportion of dual-earner couples, but few families, and a postponement of marriage and childbearing (Warde 1991). In Smith's (1996) New York study, the only census tracts studied that did not conform to this pattern were those in which gay men dominated the gentrification process.

Urban change paralleled changes in the role of women, with gentrification following the influx of women into the work force, especially the growing numbers of highly educated women in high-end service industry jobs. Many theorists (e.g. Beauregard 1986; Bondi 1991a; Markusen 1981; Rose 1984; Wekerle 1984) argued that the move of women into the labor force, and the concomitant search for urban neighborhoods that could accommodate women's dual roles at home and at work and provide the necessary services, accounted, at least in part, for the existence of gentrification. In this conception, women were not only potential beneficiaries of gentrification, but drivers of the process. Wekerle (1984: 11) argues that women were a major impetus for the revitalization of the North American city, that they both needed and gained more from an urban location since "women use the city more intensively than men and for a wider range of functions: work, childrearing, shopping, cultural facilities and neighborhood participation. Women also gain more time since they show the greatest decline in travel-to-work time after a move from the suburbs to the city." These benefits accrued to women despite the fact that the city was still very much a place planned by and for men. What would be required then, in light of the fact that these demographic and economic shifts were not going away, was a major change in the land use patterns of cities.

This change in land use patterns and gendered divisions of labor has not yet been realized. Markusen (1981) argued that just as the dominance of the single-family home, its separation from the workplace, and its distance from the urban core was as much the result of patriarchal organization as it was the result of the capitalist organization of work, gentrification marked a remaking of urban space that reflected changes in both patriarchy and capitalism. The rise of both women and gays in the managerial classes was necessary to creating the demand for more central urban locations. Two-income households were both a by-product of the rise of the new middle class (Ley 1986) and a requirement to afford the increasing costs of urban housing. While the structure of work changed with deindustrialization, urban spatial structure was slow to respond. As Markusen (1981: 33) recognized,

> its existence continues to constrain the possibilities open to women and men seeking to form new types of households and to reorder the household division of labor. The fact that housing, the primary workplace for social reproduction, is also the major asset for many people tends to reinforce the single-family, patriarchal shape of housing and neighborhoods.

These limited choices affect all women and caregivers. The inefficiency of this polarized organization of urban space, individualizing child care, travel to work, home maintenance, meal preparation and so on, increases the burdens that disproportionately fall on women and others in caring roles (Markusen 1981; Mackenzie and Rose 1983; Hayden 2002). Empirical work on the gendered practices of gentrifier households found no difference from their suburban counterparts (Bondi 1999; Mills 1988). The materiality of cities still overwhelmingly reflects the fundamentally gendered assumptions that formed them. This is evident in everything from zoning plans and housing types to, as I have written about elsewhere, snow plowing, where major streets and the city center are prioritized over the side streets where schools and daycare centers are often located (Curran 2014). Rather than contest these assumptions, gentrification has largely solidified them, further exacerbating the competition for the best urban spaces in ways that build upon the inequalities of the industrial city that gentrification is remaking.

Gender is just one of the lenses through which we can view how gentrification reinforces urban inequalities, but given how important it is to the shaping of urban space, it has been profoundly under-studied, and the work that has been done is under-cited in the literature. Attention to gender in the process of gentrification has tended to focus on the role of gentrifiers rather than on the effects of those who find themselves gentrified. This was true of the original burst of interest in the role of women in gentrification in the 1980s and in more recent work on women and gentrification in the (excellent) work of Leslie Kern (2007, 2010a, 2010b, 2013) on the way in which gendered assumptions have been reinforced in the gentrification of Toronto (see van den Berg's (2013) work on Rotterdam for a partial exception).

But while gentrification allowed access to a panoply of urban benefits for those middle class women who could afford them, it is also true that women were those most likely to be displaced by the process (Rose 1984; Bondi 1991b). Women are the most disadvantaged members of already disadvantaged communities (Jupp 2014). The polarization on the landscape that gentrification represents parallels the increasing polarization in the labor force, with women concentrated in low wage labor and the gender wage gap high globally (Grossman 2014). Women are also more likely to be single parents. Responsibility for children continues to reduce earning power in what has come to be known as the motherhood penalty (Budig et al. 2012). And the increasing population of elderly are overwhelmingly female, living on fixed incomes and vulnerable to displacement (Bondi 1991b). Failure to recognize the gendered nature of displacement is not simply an oversight; it is an obfuscation of the process of gentrification and what it is actually accomplishing on the ground.

While "gender" may be a contested and fragmented field of analysis, it undoubtedly remains a felt reality within everyday lives (Jupp 2014). My goal here is to take an intersectional approach to the relationship between gender and urban space through the specific process of gentrification in order to argue that we cannot fully understand the effects of gentrification if we are not looking at the way in which gender intersects with race, class, sexuality, immigration status, age and ability to disadvantage and displace populations now deemed undesirable in the urban core. As Bondi (1991a) has argued, the refusal to consider the centrality and complexity of gender has hampered the gentrification literature's attempt to move beyond the dichotomies of production and consumption, structure and agency, economy and culture. This does not mean an exclusive focus on women; I am interested in how gendered assumptions have served to reinforce race, class, age, and other inequalities on the urban landscape.

There is an extensive literature on gentrification, an extensive literature on women in the city, and to some extent, on gender in planning, but these literatures have too rarely been in conversation with each other. This book is an attempt to further that conversation so that we can achieve changes on the ground that realize some of the emancipatory potential once hoped for gentrification without the social, cultural, economic, and physical displacement gentrification invariably brings.

Mackenzie (1988) argued that the process of gender constitution and the process of constituting urban environments are inextricably linked. In her historic look at Canadian cities, she found that periods of urban transition coincided with periods of gender role transition. Industrialization separated the traditionally unified sphere of home and work, bringing women into public spaces in a way that was deemed dangerous. The solution, suburban homes that separated women and children from the temptations of the city, became less sustainable as women were forced into the workplace as the costs of domestic work and raising a family became ever more expensive. This dual role then became the problem (Mackenzie 1988). Gentrification was seen as a solution to the problem of the separation of work and home (Warde 1991). But as with the previous "solutions," this brought its own problems.

Gentrification offered a market-oriented, individualized, privatized spatial solution to the problem of work-life balance. With urban planning failing to catch up to the lived experiences of urban dwellers, those who could afford to found more advantageous spaces in which to attempt the balance, "redis-covering" inner city neighborhoods which offered easy access to downtown jobs and other amenities. But this privatized solution further served to dis-advantage working class women, people of color, immigrant communities, seniors, people with disabilities, and other less advantaged urban populations. Increasing demand for urban housing made it less affordable. The condo boom reduced the supply of rental housing. The transformation of public housing into mixed-income housing displaced the majority of public housing residents from the redeveloped projects. The domestic labor of women, never sufficiently valued by the market, was replaced by low-income labor, often done by working class women, many of them immigrants, in sectors like child care, house cleaning, and food preparation. Private security patrols and more aggressive policing are supposed to make gentrifiers, especially women, feel safe even as these policies criminalize people of color, especially youth. The problem of urban education leads to ever more extensive private schools and charter schools that starve the public system. Urban governance is dominated by private groups that are both profoundly undemocratic and dominated by men and masculinist agendas. Urban safe havens such as gayborhoods become commodified and end up displacing the communities they were cre-ated to serve. The inequalities created by these privatizations are the urban problems of the 21st century and the topic of this book.

While women have long organized around the struggles for housing, child care, and neighborhood preservation, these movements have often been neglected in the literature of urban social movements (e.g. Castells 1983). Organizing around gentrification offered the possibility to understand urban space as both a capitalist and patriarchal problem. Rose (1984) offered this possibility in her seminal article "Rethinking gentrification: Beyond the uneven development of Marxist urban theory." Rose rightly argued that it is essential to understand the relationships between gentrification and the reproduction of labor power and people in order to overcome the polariza-tions created by gentrification. So, for example, "gentrification by employed women with children may be a deliberately sought out *environmental solution* to a set of problems that are inherently *social problems*" (Rose 1984: 66, emphasis in original). In other words, gentrifiers are struggling with similar constraints as those they displace, though of course, with different resources to cope with these constraints. Thus, "the moderate-income women's envir-onmental solution to the problems created by her dual role exacerbates the problems of the low-income woman who is displaced to other neighborhoods which are more environmentally restrictive and less socially supportive" (Rose 1984: 66). Rose argued that we need to see gentrifiers as more than just the bearers of a process determined independently of them; we can see the needs they have in common with those with whom they compete for space. Failure to see these commonalities forestalls the political possibilities that these shared challenges could create by forging alliances between gentrifiers and those they could potentially displace. What alternatives could these alliances develop to challenge the limits of urban form?

Rose proposed this possibility over 30 years ago, and yet we have seen very little recognition of the common struggles and political potential this under-standing of gentrification could provide. As McDowell (1983: 69) recognized over 30 years ago, "[t]he majority of women in the city are defined in rela-tionship to men within a locality instead of in relation to other women in different areas." I argue, following Bondi (1991a), that it is the refusal of much of the gentrification literature to recognize the role of gender that limits our conceptions of how gentrification might be replaced by more genuine strategies for urban revitalization. Class and gender are co-constituted; gender inequality and particular visions of the masculine and feminine are "core ingredient[s] of class formation and consciousness" (Hart 1989, quoted in Bondi 1991a: 193). Bringing gender into the gentrification debates goes beyond recognizing the role of, and effects on, women, to recognize the "structures of patriarchy" (Walby 1989, quoted in Bondi 1991a: 196) to which gentrification contributes. Bondi suggests exploring the gender (as well as class, race, and ethnicity) ideologies of those involved in gentrification, from architects to those displaced, as a way forward. Her own research (Bondi 1999) indicates that gentrification is not nearly the emancipatory process once hoped. In her study of two inner-urban and one suburban neighborhood in Edinburgh, she finds no evidence of variations in gender practices. While a majority of those interviewed expressed egalitarian views about gender and employment, the raising of children challenged these aspirations towards gender equity. The environmental solution of gentrification has not solved the social problem of an inadequate infrastructure for care. Whatever remaking of urban space gentrification has accomplished, it has not even begun to address this fundamental inequality.

Gentrification has become, as Smith (2002) argued, a globalized urban strategy; it represents the leading edge of change at the urban center. That is why a thorough understanding of its consequences is so vital. This does not mean that gentrification happens the same way in every city. Place matters, and local policy matters enormously. But gentrification produces urban land-scapes that convey the same aspirations and reproduce the same inequalities (Smith 2002). The mobility of policy means that policies that accomplish gentrification can be replicated in wildly different geographic contexts (Mountz and Curran 2009). While the examples cited throughout the book are largely, though not exclusively, from North America and Europe, where gentrification research has previously been focused, the processes that have accomplished gentrification are decidedly global.

This ubiquity has led to what Schulman (2012) calls "the gentrification of the mind," in which mainstream consumerism has come to replace a more transformative radicalism (see also Paton 2014). Among both planners and academics, we are perennially seeing gentrification talked about in the same way, often by the same people, to the exclusion of other voices. We are presented the false choice between gentrification and urban decline (DeFilippis 2004), with gentrification presented as the only way to attract the necessary capital to accomplish urban upgrading. This has been enshrined in urban policy. As Schulman (2012: 27–8) argues,

> Spiritually, gentrification is the removal of the dynamic mix that defines urbanity- the familiar interaction of different kinds of people creating ideas together. Urbanity is what makes cities great, because the daily affirmation that people from other experiences are real makes innovative solutions and experiments possible. In this way, cities historically have provided acceptance, opportunity, and a place to create ideas contributing to freedom. Gentrification in the seventies, eighties, and nineties replaced urbanity with suburban values from the sixties, seventies, and eighties, so that suburban conditioning of racial and class stratification, homogeneity of consumption, mass-produced aesthetics, and familial privatization got resituated into big buildings, attached residences, and apartments. This undermines urbanity and recreates cities as centers of obedience instead of instigators of positive change.

This is particularly true with gender relations. As Kern (2010a) argues, despite the autonomy and freedom experienced by many women in the revitalized city, many of the ideologies and discourses at work reconstitute patriarchal relations and remarkably traditional views on the roles of women. Rather than gentrification reshaping the ways we think about work and home and reformulating the organization of urban space, it has intensified the pressures associated with a more traditional, suburban gendered division of labor. Home ownership and the single family home remain the idealized norm. Public transit remains underfunded and less safe for women. Public spaces are still relentlessly gendered and increasingly privatized. We are no closer to the communal strategies of the non-sexist city than we were 30 years ago (Hayden 2002; Rose 1984; Wekerle 1984).

Academic work on gentrification has allowed these silences. Gentrification work has been largely repetitive, consistently focused on the same debates, and citing the same authors in a closed citation loop that has us running in circles without greater understanding. These standards ignore gender and tend to under-cite women who write about gentrification. A recent study by the Federal Reserve of San Francisco (Zuk et al. 2015) is touted as the most comprehensive review yet of what we know about gentrification. It is 79 pages long. The word "women" does not appear once. The word "gender" appears only once, in a quotation from Wyly et al. (2010). Lees, Slater and Wyly's (2008) definitive text on gentrification devotes five pages, with some additional mentions, to the issue of gender in a manuscript that runs 277 pages. This is both a reflection and a continuation of the marginalization of women and gender in the literature. The literature on gentrification in the Global South has tended to follow the example of the literature from the Global North and ignores gender. Janoschka et al. (2014), for example, offer what they characterize as a comprehensive review of the trends in the literature on gentrification in Spain and Latin America. They do not consider the question of gender and fail to cite some of the literature that does (e.g. Wright 2004, 2014; Mountz and Curran 2009). In Lees, Shin, and López-Morales' (2015) edited collection on global gentrifications, "women" and "gender" don't make it into the index, and none of the contributions focus on gender.

As this book will demonstrate, there is plenty of evidence for gentrification as a gendered practice, but too little of this work has been incorporated into the literature. The overwhelming focus on class has been to the exclusion of other forms of oppression despite the fact that these are irrevocably linked.

But it is also still true, as Wekerle (1984) argued, that a woman's place is in the city. The centrality, connectivity, and community of the urban has long served women and other marginalized groups, and despite the fact that cities are still overwhelmingly planned by men, women have shaped the city in ways that have improved urban life (Hayden 2002; Spain 2001; Wekerle 1984). As I detail the ways in which gentrification has further entrenched gendered stereotypes and gendered divisions of labor, the goal is to look for new possibilities that allow for coalitions among urban constituencies and ultimately new public policies that allow us to move beyond gentrification as the only path to urban revitalization.

The plan of the book is as follows. Chapter 2 deals with housing. This chapter argues that the typical understanding of gentrifiers as young urban professionals who postpone childbearing has resulted in a very limited housing stock that prioritizes small units at the expense of family-friendly housing, thus limiting choices for alternative residential models. The rise of condos at the expense of rental housing has further served to limit residential choice. Female heads of households with children are those most negatively affected by this changing infrastructure, but even some gentrifiers find their housing choices limited when they start having children, leading many to move to the suburbs and thus recreate the gendered divisions that gentrification was supposed to undo. The replacement of multi-family housing with condos for singles or couples, and the displacement of those communities of families, make the multigenerational family that was the safety net in the past increasingly untenable. Senior citizens, who are predominantly female, are especially vulnerable. Gentrification has also been a driving force in the remaking of social housing, which targets a population that is overwhelmingly female. The fiscal crisis of 2008 and beyond further disadvantaged the working class in gentrified neighborhoods, with women and children at the forefront of the foreclosure crisis.

Chapter 3 focuses on gendered labor markets. Central to the literature on gender and gentrification has been the changing role of women in the labor force. This chapter will update that work by looking at the gendered division of labor in the modern city, paying particular attention to the glass ceiling, the gendered wage gap, and the growth of low paying jobs in industries that have served to replace the free labor of the stay-at-home mother, such as day care. In requiring dual incomes in order to afford the gentrified city, gentrification limits choices and displaces working class women, who then face an ever greater spatial mismatch between the jobs necessary to make gentrification work for the upper and middle classes, and the far flung neighborhoods where the working class can afford to live. Additionally, gentrification has served to displace good working class jobs in manufacturing, a loss which has been felt most acutely by working class men whose masculinity becomes "redundant" (McDowell 2003) and by immigrant women in industries like garment manufacturing. The work of planning is still overwhelmingly done by men, with little recognition of how gendered their policies are, resulting in the rise of what Tickell and Peck (1996) have termed "the Manchester men," overwhelmingly white men who dominate urban governance regimes with a masculinist, business-oriented growth agenda.

Social reproduction is the concern of Chapter 4. Gentrification has done very little to undo the dual role of women at home and at work and has rather made it much more labor intensive and expensive, with increasing expectations in what constitutes adequate preparation of children for the wider world (e.g. baby yoga, music classes, language classes, etc.) in gentrified neighborhoods. Across classes, in most countries, there is an appalling dearth of quality daycare and daycare is wildly expensive; this burden falls disproportionately on women. The changes in urban infrastructure that would facilitate the navigation of home and work responsibilities which some theorists predicted in the 1980s have not materialized. The gentrification of the school system, with the rise of charter schools and a wave of school closures, has made urban parenting infinitely more difficult. Since schools serve as one of the most central of neighborhood institutions, school closures further displace community. The move away from the democratic governance of schools, from local school councils to more centralized administration, part of a larger privatization of urban governance, disempowers the women who had been school and neighborhood activists in favor of the overwhelmingly male actors from the private sector who favor a privatized agenda that undermines democratic process. The history of urban struggle is subsumed by a narrative that equates urban improvement with a good coffee shop (de Koning 2015).

In Chapter 5, I focus on urban safety. Coding the city as feminine space has been one of the ways to sell the city as safe. And yet the way this has been done has served to reinforce gendered stereotypes, with gated communities and buildings with doormen being specifically marketed to women. The selling of the safe city has actually put women at risk, with police departments in cities across the United States downgrading crimes such as rape to make the city appear safer than it is. The safe city targets vulnerable populations such as sex workers, especially trans women. Meanwhile, policing policy serves to reinforce negative stereotypes about male bodies of color, criminalizing youth of color simply for being visible in public space. Immigrants feel more vulnerable with an increased police presence. "Safe" neighborhoods become dangerous to their long-term inhabitants (Schulman 2012: 30). The lack of affordable housing makes women more likely to stay in abusive relationships to maintain housing, or to be subject to sexual harassment from landlords and repairmen. The constant potential for displacement from gentrification makes people feel unsafe in their own neighborhoods.

Chapter 6 looks at the link between gentrification and queer spaces. While early work on gentrification and sexuality often presented gays as an important constituency for gentrification, I argue that gentrification is profoundly not queer. The selling of the gayborhood has served as a class project to whitewash what had been more transformational urban spaces, resulting in the displacement of LGBTQ people from the gayborhood. This is especially true of lesbian enclaves, given the less secure financial positon of women compared to (white) men. Queer youth of color and trans people, especially trans women, are often targeted as the increasingly wealthy gayborhood demands a larger police presence. I also explore questions about whether gentrification increases prejudice against LGBTQ residents. Overall, the effect of gentrification has been to de-queer the gayborhood. I offer some ways in which we might genuinely queer the neighborhood.

In the conclusion, I will revisit some of the more hopeful predictions for how gentrification would rework gender in the city (e.g. Rose 1984) and look at how we can potentially move forward to redo/undo some of the more negative effects of gentrification on gender norms. There are obvious connections to be made across class on issues such as child care, education, and environmental sustainability that may enable people to find common ground across what has been the very divisive issue of gentrification. The us vs. them narrative of much gentrification coverage serves to politically isolate underserved groups in a way that is strategic to achieving their displacement. Forming alliances that recognize the common oppression of gender is one way to contest gentrification as inevitable, the only path to urban revitalization.

2 Housing

Gentrification has often been seen, first and foremost, as a housing issue, with Ruth Glass' original 1964 definition focusing on housing:

> One by one, many of the working class quarters of London have been invaded by the middle classes- upper and lower. Shabby, modest mews and cottages ... have been taken over... and have become elegant, expensive residences. Larger Victorian houses, downgraded in an earlier or recent period... have been upgraded once again... The current social status and value of such dwellings are ...enormously inflated by comparison with previous levels in their neighborhoods. Once this process of 'gentrification' starts in a district, it goes on rapidly until all or most of the original working class occupiers are displaced, and the whole social character of the district is changed.
>
> (Glass 1964: xviii–xix)

Housing is a gendered, raced, classed, aged, and abled issue. Gentrification has served to highlight and exacerbate these inequalities in the housing market. Although gentrification was offered as a strategy to challenge patriarchy, accommodating women's dual roles by allowing women to live in locations closer to the city center where well-paying jobs were available, it is also women, as well as other economically vulnerable populations, who are more likely to suffer the negative consequences of gentrification, most noticeably displacement from housing in gentrifying neighborhoods (Rose 1984; Bondi 1991a; Bondi 1991b; Muñiz 1998; Kern 2010a). In their study of Chicago, Nyden et al. (2006: 35) identified women as the population that is most negatively affected by the cycle of gentrification and displacement, particularly single women with children. Between the higher poverty rate of single-parent and female-headed households and the discrimination these women continue to experience, single women with children are most at risk for displacement when housing costs begin to rise.

By increasing the cost of central city housing, gentrification has served to reinforce the dual role of women by requiring an ever greater income to afford to live in central areas of the city. This affects both the women we might call gentrifiers as well as those who are displaced or who struggle to afford to stay in the neighborhood. This chapter explores some of the ways in which the housing market and the social construction of home under gentrification are gendered in a way that further disadvantages women and others gendered as female across race, class, age, and physical ability. While the gentrification of urban housing has fundamentally changed the look of the urban landscape, it has done nothing to change the gendered organization of space or expectations of what it is that home is for. Home is still overwhelmingly understood as feminine space and the search for housing and upkeep of the home still very much women's responsibility (e.g. Davis 2004; Muñiz 1998). This is exacerbated in households with children (Miller 2015). Large scale studies have shown explicit discrimination against families with children in the U.S., especially single parents with children; Lauster and Easterbrook (2011) find that this discrimination extends to single fathers. The care of children makes both the getting and keeping of housing more challenging. A study in Baltimore found that most people who are called to rent court – and ultimately evicted – are black women living near the federal poverty line and raising at least one child. Though black women make up 34 percent of Baltimore's population, they comprised 79 percent of those surveyed in the rent court study (Cohen 2015).

These challenges are only exacerbated by the fact that as gentrification has become global, so too have housing markets. Housing markets in cities which are desirable internationally, ranging from London and Melbourne to Dallas, now serve a global consumer base, thus increasing housing costs throughout the market (e.g. Searcey and Bradsher 2015; Satow 2015). There are any number of ways in which this leads to a further gendering of the housing market. In this chapter, I will look at how global gentrification has remade the housing market in three particular ways. First, gentrification has changed the physical form of the housing market, increasing the supply of very particular types of housing (condos, luxury high rises, etc.) while decreasing the supply of housing appropriate for families and communal uses. Second, in its search for new spaces in which to realize the rent gap, policy makers and developers have targeted social housing for development, severely restricting the supply of guaranteed affordable housing in central city areas around the world. The population of social housing tends to be overwhelmingly female, with concentrations of women with children and the elderly. This is related to a third way in which gentrification operates in a gendered housing market, the targeting of the elderly for displacement, the majority of whom are women.

Housing type and tenure: or, condo boom (and bust)

Gentrification is often described as a back to city movement, as people rediscover the urban core and the amenities therein, while rejecting the monotony and long commutes of the suburbs. While Smith (1979; 1996; 2002) long ago debunked this myth, declaring instead that gentrification was a back to the city movement of capital, not people, it certainly is true that gentrification has resulted in intensified development in the urban core. This has opened up new spaces for redevelopment, for example in the rezoning of previously industrial areas for residential use (more on this in Chapter 3), as well as denser development, with a higher number of housing units than seen previously. This was exemplified by the rise of the condo boom. In cities across the globe, condos became the fastest growing segment of the housing market, with buildings that were previously rentals converting to condos, but even more visibly, new high rise developments, whether condo or luxury rentals, remaking entire neighborhoods (see Figures 2.1 and 2.2). Condominium and luxury high rise development have become emblematic of the shifting housing market in gentrifying cities, representing the individualization of the housing market, focusing on young professionals, empty nesters, or investment buyers. In many cities, condos are "the landscape face of embourgeoisement" (Ley 1996: 49).

Because the markets for these developments in cities experiencing gentrification are global, the pressure on local residents looking for housing is enormous. Even after the housing bubble burst, international buyers are putting tremendous upward pressure on local markets. The situation is particularly visible in London, where central areas dominated for centuries by British

Figure 2.1 New luxury housing on the Brooklyn waterfront

Figure 2.2 Part of the Nine Elms development, Battersea, London

aristocrats and top earning professionals in banking and law are rapidly being turned into empty "safe havens" for international capital, and contributing to a ripple of dramatic house price rises in areas previously considered undesirable (Prynn 2016). Following the global financial crisis, international buyers are looking less to the gold coasts of major cities, but, as in New York, to gentrifying neighborhoods such as the Far West Side, Long Island City, and parts of Brooklyn (see Figure 2.1), or the Nine Elms development in the Battersea area of London (Figure 2.2), attracted specifically to the fact that these neighborhoods are improving and therefore present a long-term value (Satow 2015). In Manhattan, estimates are that foreign buyers account for one third of condominium purchases and as much as one half of buyers in new developments. These buyers are not just the 1 percent; rather, they are considered middle class, focused on properties in the $3 million and below range (Satow 2015). This is affecting the affordability of housing in cities like San Francisco and New York as demand outpaces supply and bidding wars result (Searcey and Bradsher 2015). But, this is not simply confined to the most global of global cities; it is now spreading to middle America, where prices are more modest and therefore have room to grow, with international buyers often beating out local purchasers because they can pay cash and therefore require no mortgage (Searcey and Bradsher 2015). Many of these purchases serve solely as investments, with buyers seeking to capitalize on surging rents in much of the United States (Searcey and Bradsher 2015), the result of the tightened credit market that is a consequence of the previous housing bubble.

The assumptions about who these new landscapes are for are entirely gendered, as illustrated by Fincher's (2004) study of condominium developers in Melbourne, Australia. First, Fincher recognized that this development world is profoundly masculine, with work environments that are dominated by men. This continues the long history of urban planning by and for men. While there is a great deal of talk by developers about the role of the liberated woman in creating a market for condominiums (Fincher 2004; Kern 2010a), as well as other forms of gentrified housing, assumptions about what is best for these women are being made by men in very stereotypically gendered ways. These assumptions fail to recognize the complicated realities of women's lives and changing roles. Despite the widespread predictions of how women's mass entry into the labor force in the 1980s would change cities, condos are part of the reinscription onto the landscape of the same gendered attitudes about home and family that made the industrial city.

Kern (2007, 2010a) details how this played out in Toronto, in many ways a poster child for the phenomenon, but the process has been global. Home-ownership was seen as the best way to build wealth, both at the individual level and for investors globally, as international real estate became a safer and more profitable outlet for speculative capital than other options. Fincher (2004) notes the "income level principle" behind condominium development, part of an anti-rent culture, in which high income people are deemed more decent than low income people. The fact of homeownership denotes a right to citizenship, a right to the city, often denied renters. Rent dominated areas are marginalized in popular narratives, presented as blighted and unstable (Kern 2010a) and thus requiring the investment and stable community that gentrification via condo development provides. As Nast (2002: 881) argues, "in all class-riven contexts, idealistic images of home, nation, family, community, and so on are hegemonically produced in the interests of those who can afford or benefit from them."

Though women have been a major demographic in the purchase of these new condos (Rose 1984) and were often targeted as buyers for these new developments, with condos representing the independence and autonomy of the modern woman through homeownership (Kern 2010a), there are profoundly gendered assumptions behind the development of these kinds of housing. While the rise of the condominium market for young professional women is based, at least in part, on the recognition that the roles of women have changed, they assume that condo ownership is just a first step on the more traditional road to marriage and family. It allows them to be rational actors, building wealth through homeownership while not requiring too great a commitment when it proves time to move on via marriage and the assumed arrival of children. In this narrative, the condo offers a way of asserting female independence and claiming economic power without forestalling the return to a more traditional path. As one owner profiled in a *New York Times* article about the trend of "landlord as pal" explained, "Owning this place makes me feel secure as a woman...I'll never have to move back to my parents' house, I'm never going to be homeless" (quoted in Koenig 2016). Now married, this owner rented out her condo, extra income that allowed her and her husband to buy a bigger house with the arrival of their child. Thus, the liberation from the patriarchal confines of marriage and family offered by the condo is temporary (Kern 2010a).

This is part of the "ladder of life" narrative provided about condos, assuming that condos are appropriate only at certain stages in the life course, either for retirees, or the other major demographic, young professionals, including single women and couples without children, including gay couples. Fincher (2004) finds that empty nesters are just a small proportion of the market, and so most of this housing actually is being purchased by young professionals. Given the changes in the housing market and constriction of mortgage finance, these young professionals may find themselves committed to their condos for far longer than originally planned. Even before the 2008 financial collapse, Kern (2010a) found some of the condo owners she interviewed to be worried about the difficulties of paying the mortgage, especially as single women.

This makes the flawed assumptions about who the condo market serves all the more problematic. In Fincher's (2004) study, developers presented empty nesters as their target audience, presenting these households as those who had earned a reprieve from the suburban home and the responsibilities that come with it, including care for children and others. But, the assumptions about the proper way to live in these types of developments affect the ability of those who reside within and future residents to adapt to different stages of the life course. Developers' assumption that high rise condo living is almost by definition child-free is a profoundly gendered matter, though it is not presented as such (Fincher 2004). "Lack of family" is the market niche condos were designed to fill. The failure to provide any sort of amenities for families and children, from childcare to school to old age services, is therefore presented as an almost moral stance, with one developer stating, "I just don't think it's appropriate for kids to be brought up in that area" (quoted in Fincher 2004: 335).

And yet, children are being brought up in these areas, with the credit crunch keeping a whole generation of families in their starter homes (Preville 2014). Again, Toronto is at the forefront of this trend, with downtown Toronto being reshaped by a baby boom. The number of preschool-age children is rising fastest in those areas where condo towers are going up. Yet, the assumption about family-free living at the core continues, so that even as the "condo kid" phenomenon continues, developers continue to plan for small units, with even the two bedroom units getting smaller, what Preville (2014) describes as a gradual, deliberate shrinking of the standard of living in Toronto.

The preponderance of small units, studios, one- and, at most, two-bedroom apartments, presupposes this housing will be used by singles and childless couples. This limits the types of families who can live in these developments. Those with large families, whether because of children and/or multi-generational living, are at a particular disadvantage. This limited view of urban housing also shapes the rental market, with many condo units entering the rental market when owners move on to other properties without being able to sell their condos.

The continued push to build for young professionals and empty nesters, and the accompanying gendered assumptions about how they live, affect urban planning surrounding these developments, with limited amenities for children and other care work that makes the work of social reproduction that much more difficult. I will expand on this in Chapter 4. As Fincher (2004: 329) argues, it is important to interrogate and challenge these assumptions, because they help to create "the forms of housing that are imagined and therefore are built." In the Pilsen neighborhood of Chicago where I do research, the Chicago Metropolitan Agency for Planning (DPD and CMAP 2015) in its plan for 2040 calls for the construction of more high-rise, small apartments in Pilsen to serve a population they predict will be dominated by retirees and childless young professionals. This in a neighborhood that is currently dominated by Mexican-American families who often live in mul-tigenerational households. But, of course, if you only build for a certain population, it is more likely that is the population you'll get.

The celebration of homeownership ignores the fact that homeownership is rarely as readily accessible to women as it is to men due to systematic dis-advantage. Women have a lower rate of homeownership due to lower incomes, higher dependency on social assistance, concentration in social housing and the low-end rental market, and need for alternative forms of tenure, like co-op and non-profit housing (Kern 2010a), not to mention out-right discrimination. A 2015 study in the Chicago region (Woodstock Insti-tute 2015) found that mortgage applications from women across all racial categories were less likely to be originated than mortgage applications from men, even controlling for loan-to-income ratio, with female-headed applica-tions 28.3 percent less likely to result in the origination of a mortgage than for male-headed applications. This was true for both conventional and govern-ment-backed mortgages. Research also showed that women were more likely than men to have received subprime loans during the housing bubble and that this disparity was greater for women with higher incomes (Woodstock Institute 2015). So even those women most likely to be able to participate in this market will do so at a disadvantage.

This makes the stigmatization of renters all the more troubling and the more gendered. The ability to achieve homeownership has been presented not just as an economic benefit, but a moral accomplishment, with homeowners imbued with properties that are assumed to make them desirable citizens. The flip side of this story is the assumption that those who cannot behave this way are to blame for their own misfortune (Kern 2010a). The neoliberal celebration of private ownership blames renters at the same time that it limits their choices. The increasing number of units resulting from the condo boom did little to help the housing market for renters (Kern 2010a), and indeed, condo conversion of existing units further served to drive up the cost of renting in cities experiencing gentrification by constricting the supply of rental units, as rental housing was converted to condominiums, especially in gentrifying neighborhoods (Center for Urban Research and Learning et al. 2007; Klodawsky and Spector 1988).

While the physical character of gentrification when it comes to the reformulation of the housing market will vary in different geographic contexts, the fundamental reworking of home is a global process, with increasing privatization of home and the responsibilities of the home, and a discouraging of historic, collective responses to the work of home. This is the result not of individual choices by consumers, but of a large scale push by urban and national governments to create these privatized environments. Indeed, Lees et al. (2015) argue that global gentrification is now predominantly state run. Here, the experience of Seoul is illustrative.

Seoul's growth does not follow the typical gentrification narrative in which an industrial city experiences disinvestment with the onset of deindustrialization, leading to decline and the creation of a rent gap which then encourages reinvestment when the central city is "rediscovered" by capital. Rather, Seoul experienced explosive growth with the flood of refugees following the Korean War and experienced rapid, condensed urbanization under a developmental state with centralized policy focused on economic growth (Shin and Kim 2015; Ha 2015). This was enacted on the urban landscape through a series of urban development projects aimed at the wholesale upgrading of dilapidated, often illegal, traditional neighborhoods by constructing high-rise apartment blocks, promoting "vertical accumulation," the goal of which was to maximize accumulation through high-density construction. This has become the defining feature of Korea's urbanization and has helped to feed Seoul's highly speculative real estate market (Shin and Kim 2015; Shin 2009), for Seoul houses about 20 percent of the national population on 1 percent of its land mass (Shin and Kim 2015).

Originally focused on the urban fringe in the 1970s, the redevelopment programs drew so much interest from both home buyers of the new middle class and speculators that substandard housing in Seoul's more central areas was targeted. What followed in the 1980s was the Joint Redevelopment Program (JRP) in which joint contributions from property owners and builders supplied development finance to build and market the new developments. Property owners who could not afford their share of redevelopment expenses could sell their right to buy a new unit at construction cost to speculators. There was no voice for tenants. This occurred in areas in which most residents lacked official ownership and upgrading was prohibited. The JRP allowed for the realization of a tremendous rent gap, at a time when Seoul was about to host the Olympic Games.

This policy was accompanied in the 1990s by a densification of housing in existing low-rise neighborhoods. Planning regulations were relaxed to allow for the conversion of detached one- to two-story dwellings into three- to four-story buildings, with the additional units to be rented out. Shin and Kim (2015) estimate that 750,000 units were created in this way from 1990 to 2001, accounting for 66 percent of the housing units created in Seoul during this period. Medium-rise apartments were also densified. This was an especially popular method of redevelopment following the 1997–98 Asian financial crisis. Given the fairly large scale of these redevelopments, major firms associated with conglomerates like Hyundai, Samsung, and Daewoo were the most sought after brand names for apartment owners (Shin and Kim 2015).

Beginning in 2002, the New Town Programme (NTP) targeted low-rise, high-density neighborhoods for redevelopment. The goal was to bring outdated residential neighborhoods and urban facilities together into new urban districts with a total of 35 new mega districts designated, subsuming 372 sub-districts. Part of the goal of the redevelopment, which would occur through demolition and reconstruction, was to reduce the number of renter households from the existing 69 percent to 19.2 percent upon completion (Shin and Kim 2015).

Seoul's redevelopment program has focused on the maximization of landlord profits rather than improving the welfare of low-income tenants or revitalizing community (Ha 2015). State policy has focused on physical improvements with few benefits for the poor; in other words, the state adopted gentrification as urban renewal strategy (Kyung 2011). While historically in Seoul the language of gentrification is "missing from the urban vocabulary" (Ley and Teo 2014), the process was and is firmly in place (Shin and Kim 2015) as an increasing body of research shows. This large scale upgrading of the urban landscape has accomplished the displacement that is central to gentrification, and as in the Anglo-American context, has done so in a way that is highly gendered.

These new urban forms have a number of negative impacts. The gap between the rich and poor in housing conditions has deepened. The ratio of re-housing in place for original residents under the joint redevelopment program (JRP) was very low, with survey work suggesting that nearly 80 percent of the original residents were displaced (Ha 2015; Shin and Kim 2015). Similarly, under the NTP, the proportion of small and inexpensive units for low-income families having decreased dramatically (Ha 2015). This likely affects women especially, as the responsibility for housing in Korea is traditionally seen as the woman's responsibility (Davis 2004) and the income gap between men and women is greater in Korea than in any other country for which data is available (Grossman 2014).

The burden of resistance to these measures also largely fell on women. In the face of these dramatic urban upheavals, it was up to women to decide whether to move or to stay and fight (Davis 2004). As is often the case with public housekeeping, women in Seoul were far more visible in neighborhood struggles than in other social movements, though even this visibility was tempered by patriarchy. Women on Tenants' Committees reported that while women were the majority of members, responsible for the day to day organizing, turnout, posters, education and consciousness raising, men would sign off as the titular heads of organizations and would negotiate with male government officials (Davis 2004). But when it came to confrontations, women were often pushed to the front lines in the hope that a more feminine appearance would minimize the risk of the violence that was often deployed to clear existing settlements; women sustained injuries (Davis 2004; Ha 2015).

Displacement is not just physical but social. Residents of old communities reported strong community feelings and a community-based culture that was disrupted by the redevelopment. Ha (2015) conducted a survey of Gileum new town residents and found the vast majority, 67.3 percent, answered that the redevelopment had negatively affected their sense of community. While Ha does not break down these responses by gender, the experience is clearly gendered. Because of traditionally gendered responsibilities, women are more likely to be home during the day for child care and care of the elderly and therefore to be more aware of, and reliant on, neighborhood networks (Davis 2004). In Davis' (2004) study of Tenants' Committees, the women involved saw their allegiance as first to the family and then to the community. They identified with place rather than class, and while in some instances they were able to preserve their place in space, the neighborhood change to middle class caused social disruption of what had been collective experiences like afterschool centers, community production cooperatives, and mothers' associations, which functioned as places both to receive services and become politicized.

These new physical spaces reworked social norms in ways that were distinctly disadvantageous to women. The redevelopment programs in Seoul, from the JRP to the NTP, focused on apartment living which became attractive not just as places to live but as investment opportunities. High rise apartments became the most popular housing form in Korea (Ha 2015). They were seen as more modern, efficient, and safe, with features like doors that locked, indoor baths and toilets, and heat (Davis 2004). As in cities such as London, Vancouver and New York, the increasing number of women in the labor force helped to fuel the demand for new urban housing, though the rate of women's participation in the labor force is lower in Seoul than in these other cities (Kim 2006). And as in other cities, factors such as urban centrality, reduced commuting times, lifestyle, and educational environment were cited as reasons for choosing housing in the urban core (Kim 2006). For working women, the additional burden of housework and childcare made reducing commuting time an even more important factor in housing choice (Park et al 2010). While these new forms may have allowed some upper and middle class women to ease their dual burdens through housing choice, they did nothing to change the fundamental inequalities that make this choice necessary and further served to isolate women.

As in other cities, this housing redevelopment in Seoul increased the number of housing units, but for generally smaller households. Shin and Kim (2015) cite a 303 percent increase in the number of dwellings due to redevelopment in 65 JRP districts from 1990 to 1996 but only a 32 percent increase in the number of residents. More spacious single family units became the norm in neighborhoods that had previously experienced house sharing (and often overcrowding). This may well benefit individuals, but undermines the neighborhood and the sense of the collective.

These massive structures literally and figuratively separated people from the street, obscuring visual access to the surrounding terrain and disrupting local street patterns, and therefore access to the surrounding community (Ha 2015). This new form of housing disrupted traditional multi-generational neighborhoods with a close knit culture of helping one another (Davis 2004). As Davis (2004) argues, the ideology of individuated time that these apartments in Seoul represent perpetuates a system of domination founded on the inequality between the sexes. Upscale apartments simply expanded the spectrum of work women were expected to take on, while removing them from traditional networks of mutual assistance. The women of the Tenants' Committees Davis (2004) studied envision community as collective spaces, with residents at different stages of the life course living close together to lean on each other. The goal of the Tenants' Committees was to achieve low-income rentals for long-term residents and to fight the spatial separation of work and home through institutions like worker cooperatives that would allow for what Davis calls "creative production" that blurs the lines between production and social reproduction, producing both products and people, bringing the domestic into the public sphere. But accomplishing the gentrification of urban neighborhoods through privatized high-rise living, in Seoul as in other cities, forestalls these alternative ways of living

Privatizing social housing

Real estate pressure is such in some cities that the frontier of profitability has extended to social housing estates, areas previously seen as off limits for any sort of development. This move to absorb social housing into the private market under the guise of social mix has been strongest in the U.S., U.K., and the Netherlands, but is also in force in Ireland, France and Australia. Welfare programs like social housing have acted to mediate some of the worst excesses of gentrification, a luxury not available in much of the Global South, but now these too have been neoliberalized, putting the most vulnerable at risk (Bridge et al. 2012). In this remaking of the urban landscape, the role of the state is to reinforce and harden the effect of market forces rather than trying to counter them (Marcuse 1997).

Because of their relative economic disadvantage across geographies, social housing plays a greater role in housing for women (Smith 2005). Neoliberal policies have further "embounded," women, isolating them when the state revamps these policies while also withdrawing resources (Miranne 2000). While the policies of social housing reform and social mix are presented as gender neutral, they are not (Miranne 2000). In the U.S., women were already isolated in public housing because housing authorities limited or removed benefits if an able-bodied man were present in the household. This accomplished social housing as a space for women and children (Pruitt-Igoe Myth 2011). In the U.S., for example, female-headed households comprise 75 percent of public housing, 75 percent of project-based Section 8 housing, and 83 percent of housing vouchers (National Low Income Housing Coalition 2012). In Chicago, whose Plan for Transformation (PFT) to remake public housing is the most ambitious in the U.S., 88.9 percent of households in subsidized housing were female-headed (Chaskin et al. 2012). So any attempt at social housing reform is by definition gendered.

A precursor to the remaking of social housing is the pathologizing of those who live in it (Lipman 2012) and delegitimizing of the programs that serve them (Bennett et al. 2006; Chaskin and Joseph 2015). Public housing and related programs have been under attack for decades. Data analysis shows that the beneficial effects of housing vouchers and related programs, especially for single mothers, have been systematically undervalued (Meyer and Mittag 2015). Despite the fact that housing assistance is successful when it is available, and that public housing is one of the ways the working class can stay in the central city (Wyly et al. 2010), these programs are not being made available to all who need them and are being actively un-done in the gentrifying city.

That the removal of existing populations is the goal of these "social mix" policies is clear from the data. Chaskin and Joseph (2015) call mixed-income public housing redevelopment a kind of planned gentrification. Wyly and Hammel's (1999) multi-city study found the HOPE VI program that funded public housing redevelopment in the U.S. proved to be a catalyst for gentrification. In cities from Chicago (Lipman 2012) to Dublin (Hearne 2014) to Melbourne (Shaw 2012) to Seoul (Ha 2015; Shin and Kim 2015), the rate of re-housing for original residents was a mere 20 percent. The HOPE VI program in the U.S. resulted in the displacement of 200,000 units over 20 years, with only 50,000 replacement units build (Smith 2015). And across these geographies, there is no accounting for the loss of more family-friendly housing sizes in favor of housing for singles in the redevelopments. In Chicago (Figure 2.3), 14,000 units, the majority of those planned for demolition, were family units (Smith 2006). There are no reliable statistics on the number of bedrooms lost, but an increase in the number of housing units has not meant an increase in the number of people housed (Shaw 2012).

While not at the forefront of the research on social housing reform or gentrification outside of Ireland, Dublin's attempt at "regeneration" of its social housing illustrates all the flaws with privatization, the negative effects of which fall particularly hard on lone parents, the unemployed, and low-income households, with care services being de-prioritized (Barry 2014). In Dublin's inner city, 50 percent of households with dependent children are headed by a

Figure 2.3 Vacated Cabrini Green townhouses with new construction and Hancock
        Tower in the background, Chicago

single parent; since 2006 the single parent family has become the dominant
family type within Dublin's inner city (Haase 2009). As in other cities
engaged in social housing "reform," these families must deal with an affordable housing crisis resulting from privatized public housing that substantially
reduced the number of units, as well as a financial collapse that froze the
construction of new units.

During the boom years of the 1990s and early 2000s, Dublin saw rapid
growth, with 150,000 new jobs, but a lack of appropriate housing for these new
workers, leading to an affordability crisis that accelerated gentrification in

Dublin (Williams and Redmond 2014). This demand led to a building boom which was concentrated on owner-occupied housing, even within social housing. The Dublin City Council managed 25,000 units, but engaged in a series of public-private partnerships that led to a speculative land transfer of high value public land to private developers (Hearne 2014). These developments prioritized private residential and commercial use in order to attract high income users. The state actively engaged in "de-tenanting" the working class in order to make the projects more attractive to developers (Hearne and Redmond 2014). As in Chicago and elsewhere, policy makers assumed that, "mixed tenure will in some automatic way deliver socially virtuous outcomes" despite the existing literature to the contrary (Hearne and Redmond 2014: 221). By 2008, at least 12 developments had been privatized this way. While the new developments did provide some social housing, it was far below previous levels. As in the other examples of privatized public housing mentioned above, only 21 percent of the original population in 10 communities remained in place in 2013 (Hearne 2014).

As with the rest of the housing market, the 2008 financial collapse hit these developments hard. Developments that had been de-tenanted and partially demolished to make way for new construction, such as O'Devaney Gardens in the Stoneybatter neighborhood, sit empty, with no new construction underway (see Figure 2.4), while other projects slated for redevelopment saw

Figure 2.4 O'Devaney Gardens, social housing awaiting redevelopment in Stoneybatter, Dublin

a steep decline in maintenance that led to a downward spiral in living conditions, as in Dolphin House, where mold and sewage were routine factors of daily life, until female residents organized to change it (Beattie 2014; Hearne and Redmond 2014).

In a city dominated by single heads of households with children, the burden of these precarious housing situations falls on women who are expected to provide care in an age of austerity in which care services from the state have been withdrawn. Gentrification opened up new spaces for development and the realization of the rent gap with the privatization of social housing and then left those displaced or otherwise disadvantaged by the process with few resources to cope with its effects in a declining economy, with subsidies going to banks instead of citizens. This is a global process, but its gendered effects in Ireland have been particularly troubling. The austerity required by neoliberalism has resulted in cuts in services, particularly around care, that are being replaced at the household and community level by women's unpaid labor (Barry 2014). The infrastructure of equality that Ireland had constructed prior to the collapse is being dismantled, as budgets of key agencies that were focused on women, equality, poverty, and racism were slashed and important organizations closed down or absorbed into government departments (Barry 2014).

I would argue that social housing was part of the infrastructure of equality, however flawed, and that its removal through privatization as planned gentrification, in cities across the globe, has specifically targeted the women, children, and elderly who were its prime beneficiaries. Structural inequalities have accomplished the gendered segregation of social housing. The gentrification of social housing is therefore by definition gendered, as a specific population is targeted for displacement, with no viable means to access affordable housing at the urban core.

But just as the policy has been gendered, so has the resistance. The most widely known example of this is the case of the Focus E15 Mothers in London, a group of young mothers who started organizing for affordable housing when they were told they would have to leave the Focus E15 hostel because Newham Council was withdrawing funding in August 2013 (Stone 2015). Leaving the hostel would have meant leaving London; there was no other affordable housing available in the city. So they organized to stay in London. Their experience fighting their eviction led them to make connections with others experiencing the same thing. The Focus E15 campaign is a direct action campaign that tries to keep people in their homes or find them other appropriate, affordable housing. Having successfully won their own battle for rehousing in the area, Focus E15 organized the occupation of Carpenters Estate, a housing estate being kept empty in preparation for redevelopment next to Olympic Park, an area that has experienced rapid gentrification in the run-up and aftermath of the London Olympics in 2012. The occupation achieved an agreement to allow for the rehousing of 40 units on the estate. The Focus E15 Mothers keep a weekly street stall on Stratford

Broadway so that people can find them and grow the campaign to achieve "social housing, not social cleansing" (FocusE15.org). This has evolved to the opening of an office space, called Sylvia's corner (after suffragist Sylvia Pankhurst), part of the continued movement to "hold spaces in a city that private developers and landlords are trying to claim solely as their own" (Focus E15 2016c).

While the E15 movement has evolved beyond that original group of mothers, with a name change from Focus E15 Mothers to Focus E15 to reflect this widening scope, including activists beyond mothers, it is significant that the movement started that way. The movement is based on defending the right of young mothers and their children to stay in London to maintain their support networks. One E15 mother describes the campaign explicitly as "coming together as mothers" (Focus E15 2016b). The emphasis on home, support, and care has strong feminist and mothering elements, so it is unsurprising that it has a largely, though by no means exclusively, female support base (Watt 2016). As the previous discussion has illustrated, the burden for affordable housing falls disproportionately on single mothers and others involved in care work. Those with children are especially vulnerable but also, perhaps, especially sympathetic. As one Focus E15 mother commented (quoted in Watt 2016: 308):

> I think that is another reason why this campaign got quite far because a very big visual is seeing a girl with a buggy and like us all, we all looked quite military with our buggies like big tanks or something. So yes I think that is something that made it a little bit more powerful.

The fact of motherhood, the burden of care in this example, highlights the disparity between the luxury housing explosion in global cities such as London, and the everyday work of social reproduction engaged in by working class mothers and others. As another Focus E15 mother recognized, "They're not expanding and investing in the community for the people of the communities sake: it's for the sake of tourists and money!" (Focus E15 2016a). Focus E15 has emerged as a powerful force that city government must pay attention to, and has helped to provide support for other housing movements (Watt 2016). This is one example of gender and care as organizing principles to contest gentrification across geography, race, and class. The ambitions of the movement are clear. As one banner declared "this is the beginning of the end of the housing crisis" (quoted in Watt 2016: 315)

The aging city

The documentary *BOOM: The Sound of Eviction*, about gentrification during the tech bubble in the late 1990s in San Francisco, recounts the story of Lola McKay, a life-long San Francisco resident who died fighting eviction from her apartment after it had been bought by speculators. She was 83 years old.

Apartment buildings like Lola's, full of long-term tenants, can be attractive to developers because they tend to be cheaper to buy than other buildings. These long-term residents tend to be senior citizens. Many have no family. Once displaced, they most often have to move to nursing homes far from the city center. Ted Gullickson, an organizer with the San Francisco Tenants Union interviewed for the film, said that of the seniors they follow, few last longer than a year in their new home. He concludes, "The eviction of seniors kills them" (BOOM: The Sound of Eviction 2002).

The relative invisibility of older people in urban research (Buffel and Phillipson 2016) is especially acute in gentrification research, a gaping hole given that the elderly are one of the most vulnerable groups to this process. The elderly face unique dangers that extend beyond displacement and influence mental and social well-being (Petrovic 2008; Nyden 2006; Henig 1984). What research does exist shows that gentrification specifically targets the elderly, with gentrifying communities demonstrating a significant loss of senior citizens aged sixty-five or older because half of the elderly population earn less than half of their area's median income, and one-third of these elderly residents sacrifice more than half of that limited and often fixed income to pay for housing (Petrovic 2008). One study revealed the elderly occupy over half of the public housing in the United States (Petrovic 2008). Seniors are more likely to be reliant on fixed incomes, with few alternatives for increasing income to deal with increasing housing costs (Petrovic 2008; Nyden 2006). The prohibitive costs of real estate, especially in global cities, limits their choices and often forces them to move to senior developments (Portacolone and Halpern 2014; Buffel and Phillipson 2016).

The vulnerability of older adults is further exacerbated by race and gender. In the U.S., African-Americans, followed by Latinxs, experience greater segregation as a result of unmet housing needs (Petrovic 2008). And these disadvantages will be disproportionately experienced by older women, who outnumber older men. This difference widens as the population ages; in 2010, 57 percent of all adults age 65 and over were women while two-thirds of Americans age 85 and over were women (Institute on Aging n.d.). Older women are twice as likely as older men to live alone (37 percent and 19 percent, respectively). In 2010, 72 percent of older men lived with a spouse, only 42 percent of older women did (Institute on Aging, n.d.). Single women, with a median age of 78, account for 58 percent of households spending more than half their income on rent (Kinney 2016). The poverty rate was higher among women aged 65 and older than men in this age group, and higher among Latinxs and African Americans (Cubanski et al. 2015).

The burden of care for older adults, and thus for dealing with the negative consequences on those displaced or otherwise affected by gentrification, is also highly gendered. Upwards of 75 percent of all caregivers are women. In the U.S., the average caregiver is a married woman, age 46, working outside the home for $35,000 annually. These female caregivers may spend as much as 50 percent more time providing care than men and may suffer financial consequences as a result of that care; a common scenario is an older woman who cares for her husband only to discover later that there are few financial or other resources to meet her own needs for assistance. One 1998 study found that half of Baby Boomer women caregivers suffered "financial hardship" as a result of their caregiving. Meanwhile, the economic value of the informal care provided by women is estimated at somewhere between $148 billion and $188 billion annually (Institute on Aging n.d.).

The spatial concentration of seniors in disinvested neighborhoods close to the urban core makes them a particular target for gentrification (Henig 1984; Petrovic 2008). This can leave seniors trapped in homes that are no longer appropriate for aging bodies, displaced to far flung neighborhoods away from existing social networks, relocated to nursing homes, or homeless (Leland 2015b) and may even lead to premature death due to the stress and disorientation that can result from relocation. Longer life spans and systematic economic disadvantage again make women prominent among this targeted group.

And they are indeed targeted, displaced not as an accidental effect of the housing market but chosen because of their vulnerability, as Blanco and Subirats (2008) find in their study of five Spanish cities. Galcanova and Sykovova (2015) similarly find that gentrification in Prague has led to the displacement of seniors, sometimes by illegal means, and thus to intense feelings of insecurity. Petrovic (2008) details how seniors are routinely targeted for stepped up enforcement of the housing code as a way to displace them from long tenancies; cities take building codes and zoning policies designed to shield poor residents and instead use those laws as a sword to force them out. Seniors are also a target for predatory lenders. Lenders and developers will seek elderly residents out, manipulate and pressure them into selling their homes, and then take their property at unjustifiably low costs. Some developers do not even try to persuade elderly residents, instead filing code violations that incur expensive repair costs. Consequently, the elderly often sell their homes at unfair values and find themselves in similar if not more untenable situations (Petrovic 2008; Nyden 2006). This strategy is on display in the documentary *Flag Wars*, about gentrification in one Columbus, Ohio neighborhood, in which the enforcement of building codes is used as a tool to displace long-term residents from historic homes now valued by gentrifiers. Even for those who are able to remain, gentrification erects barriers, displacing existing businesses and, as such, making it more difficult to access affordable medical care, groceries, and jobs (Petrovic 2008).

This "vicious cycle" between booming housing process and the vulnerability and displacement of seniors is "greater in New York than anywhere else," according to Donna Cerrado, the commissioner of the city's Department for the Aging (quoted in Leland 2015b). Wyly et al. (2010) found a strong correlation between age and the occurrence of displacement, with older women at the highest risk. In New York, seniors pay the highest share of their income on rent: 58 percent for elderly renters compared with 34 percent for average households. Older tenants with low rents are primary targets for unscrupulous landlords who want them to move out so that the landlord can raise rents (Leland 2015b) Even those who own their homes face the prospect of rising real estate taxes that make the property unaffordable (Leland 2015b; see also Curran and Hague 2006). Seniors end up in nursing homes not by choice but because they cannot afford their housing, and there is no other available affordable housing. Some do not even have that option. In New York, about 2,000 people age 60 and over stay in the city's homeless shelters (Leland 2015b). Over half of the elderly population in New York City are immigrants, as are 40 percent of those 85 and older. These immigrants face particular financial challenges, as they may not have access to sufficient Social Security or pension funds. Family used to do this care work, but in a changing economy more people are more vulnerable. Gentrification takes place now in an era of austerity, with real estate prices booming even as social service funds are shrinking. Senior services and housing for the elderly are routinely underfunded.

A series on the oldest old in the *New York Times* (Leland 2015a; 2015b) detailed the multiple displacements and disorientations experienced by elderly New Yorkers made victim to the vagaries of the global real estate market. One such story is that of Ruth Willig who, along with more than 100 others, was told she had to leave her assisted-living home in Park Slope, Brooklyn, a thoroughly gentrified neighborhood that has served as an example of how the process occurs (Carpenter and Lees 1995). Ms. Willig relocated to another assisted living home across the borough, resulting in a host of difficulties with doing things she had previously taken for granted, namely accessing her medical appointments and maintaining her social networks. Of the experience, she said, "I find that I'm sad most of the time. I'm not happy anymore ... People in their 90s shouldn't have to be moved because this guy is greedy" (quoted in Leland 2015a). Nearly a year after her move, she remained uninterested in her new neighborhood and still missed her friends from the old neighborhood, some of whom had died since the move, deaths she was convinced were hastened by the trauma of having to move (Leland 2015a). Experiences such as these led the Crown Heights Tenants' Union, a tenants' rights organization in the Crown Heights neighborhood of Brooklyn which has been seen as the last frontier of affordability in hyper-gentrified Brooklyn, to include protections for senior citizens from displacement in their list of demands (http://www.crownheightstenantunion.org/our-demands).

These types of displacements can be contested. Cho and Kim (2016) detail the case of Jangsu Village in Seoul, in which the preservation of the existing senior population was the motivation behind an alternative neighborhood redevelopment plan that preserved and redeveloped this historic neighborhood rather than allow its demolition for luxury high rise development, as in other areas of Seoul discussed earlier. As in other global cities, Seoul has experienced a rapid aging of the population; the percentage of the population over the age of 65 doubled from 2000 to 2010 (Cho and Kim 2016). Rather than allow for the kind of large-scale urban renewal that has remade Seoul through the NTP, community members, with the help of the Alternative Regeneration Research Team (ARRT), developed an alternative plan allowing for the repair and reuse of existing buildings while fostering community. With city government funding, this has resulted in upgraded buildings that have often employed area residents, a village museum, a cooperative dining room, a café, and a community center which has been particularly used by older women. Cho and Kim (2016: 113) therefore suggest that

> the notion of aging in place might provide an alternative framework for urban regeneration – one that places more emphasis on the continuous residency of inhabitants, progressive transformation and repair, recycling and the reuse of built forms, as opposed to one large-scale urban renewal project driven by property investment.

An age-friendly city fosters intergenerational dependency, allowing for reciprocity between heterogeneous groups while building solidarity among homogeneous groups (Cho and Kim 2016). This alternative vision for redevelopment requires that we see older people as "normal urbanites" and consider their needs and lifestyles in city development. The participation of older residents is key to neighborhood governance and to making the city more age-friendly (Cho and Kim 2016: 113). And it may serve as a strategy to slow and challenge gentrification.

Conclusion

Too often the focus on the class change that gentrification brings about obscures the ways in which the displacements accomplished are gendered. In this chapter, I highlighted three examples of changes to the housing market that have particularly gendered effects. The condo boom has remade the housing market in ways that build upon longstanding gendered assumptions about the gendered division of labor and the life cycle. It has reduced access to rental accommodation and thus made what rentals are available more expensive. Housing for families has also become more expensive and difficult to access. The burden for finding housing is gendered, given that this responsibility often falls to women. The changes in the private market are exacerbated by the shrinking of the supply of social housing. While not often addressed as a women's issue, the majority of social housing tenants are women, and therefore they are the ones most disadvantaged by the privatization of social housing and the shrinking supply of social housing units. They are further disadvantaged by having to navigate the housing market with young children, and are likely to face discrimination in the housing market as a result of the burden of care. Another group targeted by gentrification are the elderly, the majority of whom are women. Their fixed incomes and concentration near the urban core make them attractive as targets for displacement and realization of the rent gap. To overlook the gendered nature of these displacements is to minimize and normalize the patriarchal systems that make these women more vulnerable.

But gender can also be a source of solidarity and community building, as the Focus E15 mothers have shown. This group of young mothers have served as inspiration for housing movements across the UK and beyond. Rethinking housing as a communal good, a fundamental aspect of care, serves to contest the inevitability of gentrification. Similarly, the example of Jangsu Village in Seoul demonstrated that successful alternatives of community development are possible when all members of the community, especially the elderly, are taken into account. So alternatives may be possible, but to achieve them we have to recognize that certain populations are marginalized in ways that go beyond, and intersect with, class.

Across class, race, age, and family status, women are more vulnerable to displacement because of discrimination in the housing and mortgage markets, which is in part an outgrowth of the gendered wage gap and other forms of discrimination in the labor market, to which I turn next.

3 Labor

Employment restructuring was central to the emergence of gentrification, with the shift from an industrial to post-industrial economy crucial to the creation both of the rent gap (Smith 1996) and the new middle class who served as the consumers of gentrified urban landscapes (Ley 1996). Changing roles for women in the labor market were cited as one of the drivers of gentrification. Warde (1991: 229) went so far as to argue "that aspects of change in the nature of women's labour-market participation account for the incidence of gentrification." In the 1980s, women had access to higher level professional jobs in the urban core as never before, and many chose central city residential locations to be close to these jobs and the other amenities an urban location had to offer. Women employed downtown were more likely than their male counterparts to live downtown (Rose 1989; Ley 1996). But rather than a remaking of the patriarchal family structure, as Markusen (1981) and others (e.g. Ley 1996) suggested, this residential choice was an attempt to reconcile increasing work demands with domestic responsibilities in a division of labor that was still highly gendered. It also exacerbated the disadvantages of working class people. The concentration of high-income white collar workers in the urban core resulted in the displacement of lower-income workers, lengthening their journeys to work in the city center (Ley 1996).

This chapter argues that although gentrification coincided with a rise in the number of women in high-end service sector jobs, the increasing number of women in the workforce was as much the result of economic necessity as feminist liberation. The gentrification of the city that was hypothesized as a strategy to undermine patriarchy has done little to dismantle the gendered division of labor, employment discrimination and the wage gap. Indeed, the coping strategies of professional women (and others) trapped in the increasingly competitive and precarious high-end labor markets that are the hallmark of gentrification exacerbate the disadvantages of the poor, racialized women whose low wages serve to replace the labor women in the home used to provide for free. Women are more likely than men to be poor in cities. In the U.S., for example, women make up two-thirds of the U.S. minimum wage work force. Women's overall burdens of paid and unpaid work have intensified, even as some women's lives improved (Parker 2016a). While coding urban space as feminine has been a useful strategy to accomplish gentrification (van den Berg 2012), norms and expectations in the labor market of the gentrified city remain gendered and highly masculinist (Parker 2008).

The work of gender in the gentrified city

The central labor market transformation that predicated the existence of gentrification was the decline of industry and the shift to service sector employment. Industrial labor was coded as hard, dirty, working class, and male. The service sector is seen as clean, educated, upper class, and more female. In attempting to remake the urban core, cities have played with these masculinized and feminized representations of work to make cities more attractive for investment (see Kern 2013; Patch 2008; van den Berg 2012).

This shift to the more "feminine" professions is one way in which the new economy created what Linda McDowell (2003) called "redundant masculinities." Since work has been so essential to the construction of masculinity, a change in the nature of work changes the nature of masculinity. The rise of service sector work, work that, especially at the bottom end, requires care and deference, is seen as "innately" feminine, so that women are therefore the preferred employees (McDowell 2003). While this could theoretically have led to a decline of patriarchy, McDowell (2003) offers that what we are seeing instead, in a period of deep anxiety about gender identity and relations, is the strengthening of a hegemonic vision of masculinity that is misogynistic as well as deeply ambivalent about race and sexuality, thus deepening divisions between women and men and among men. This can result in an exaggerated masculinity of working class men that can disadvantage them in the labor market (and as life partners for women). Rather than a crisis of masculinity, we see an uneven challenge to the automatic association between masculinity and privilege, with race and class key parts of the way masculinity is constructed (McDowell 2003).

Van den Berg (2012) employs the example of Rotterdam in the Netherlands as an example of a city strategy to create a different gender identity for this industrial city as a way to change the city's class position. Rotterdam is a city built on its industrial history centered on the city's harbor. The city's image has been that of "musclemen," where "shirts are sold with their sleeves already rolled up" (quoted in van den Berg 2012: 158). But, declares one promotional book on Rotterdam, "The city no longer has such a need for musclemen" (quoted in van den Berg 2012: 158). This reimagining of the city's image discursively produces space for a new, middle class population in more feminized professions (van den Berg 2012). Rotterdam, in its attempt to attract dual-income middle-class families, aspires to a new economy in which the lost jobs for men in the harbor and industry will be replaced by jobs for men and women in tourism, healthcare, and creative industries. "Blue-collar workers are to be replaced by pink-collar workers; masculine 'work' by, slightly exaggerated, feminine 'professions'" (van den Berg 2012: 154). Thus, van den Berg offers the term "genderfication" to describe this strategy of gentrification, where "space is not only produced for more affluent users, but also for specific gender relations" (van den Berg 2012: 164), another way in which to frame who belongs and who does not. Feminizing this transition is an attempt to mask its brutality.

This vision of femininity is very white. It is a way of accomplishing the erasure of working class men of color, especially in Rotterdam, the most ethnically diverse city in the Netherlands, as well as alternative visions of the feminine (van den Berg 2012). The eroticized vision of the city as femme fatale, as temptress (the image if the pink stiletto was used in marketing materials), invokes the hyperfeminine to both break the mythology of the hypermasculine and exclude Muslim women and others who are too traditional or not yet modern enough to conform to this new city image. It also firmly places the feminine in the role of consumer (van den Berg 2012).

Gendering city image differently is just one tool cities have used to accomplish the deindustrialization of the city in an attempt to become "global" and attract the cosmopolitan elite. My work on the effect of gentrification on small scale manufacturers in New York City finds that the removal of industry from areas of the city considered newly desirable was city policy, part of the city's rebranding as global city (Curran and Hanson 2005). Through large scale rezoning of industrial areas to individual zoning variances and a more quotidian policy of harassment with recurrent citations for things like noise, smells, and minor building code violations, city policy explicitly seeks to remove industrial land uses to open up new areas of the city for speculative real estate development, despite the fact that the urban core remains a good place to do business for manufacturers (Curran 2004; Curran and Hanson 2005; Curran 2007; Curran 2010). While often coded as male, industrial employment has been an important source of employment for women, particularly immigrant women, working in sectors such as food production and textiles. The wholesale removal of the spaces in which these industries can locate fundamentally reworks the nature of work and employment in the urban core and jeopardizes the ability of the working class to stay in what were historically port of entry communities.

Part of accomplishing the gentrification of previously working class areas of the city has been to make this blue collar work invisible while showcasing and prioritizing the more "creative" industries associated with gentrification. As Kern (2013) notes in her work in Toronto, making female workers visible has been part of this process. Female owned businesses have been part of the "hipification" of the Junction neighborhood in which Kern conducted her research, with 25–30 percent of retail businesses owned by women. While these female business owners and their "quirky" businesses help to prime the area for gentrification, they are far less likely to be adequately capitalized, and thus may help to facilitate their own displacement as better capitalized businesses, including chain stores, squeeze out independent retailers and service providers. Gentrification is part of the structural production of precarity, exposing gendered vulnerabilities and insecurities that are produced, in part, by the feminized consumption landscape, and by the inequalities that are exacerbated among women as consumers and workers in the new economy (Kern 2013).

The real estate market that feeds gentrification is itself both very masculine and masculinist (Brownill 2000). Rachel Weber (2015) in her analysis of the millennial property boom in Chicago, details the role of sociability and affect in the culture of real estate professionals. Weber (2015: 66) describes the culture of commercial brokerage houses as resembling "that of college fraternities" in which brokers have to affect the right "jocular" attitude to participate. The "discourse of the field relies heavily on sports metaphors ('slam dunks,' 'pitches'), competitive gamesmanship ('let's *do* this thing' being a commonly heard battle cry), and deal consummation to construct competition in what is a cutthroat, winner-take-all field" (Weber 2015: 67). Elements of due diligence like obtaining tenant commitment through pre-leasing "is for sissies" (p.101). LaSalle Bank was the only hometown bank in Chicago "cool enough to play with the New York boys" (p. 141). "It was like a college football game here" with teams of brokers and developers picking off tenants from their competitors and then "high fiving" (p. 164).

The work of urban planning and policy making also remains largely male and masculinist, a fact male policymakers refuse to acknowledge as important. Parker (2016a) recalls an incident in her fieldwork researching urban redevelopment in Milwaukee: "I gingerly note that there are no women on the city's 15-member council. 'Honestly ... it doesn't make a bit of difference. Whoever does the job best is what matters', he replies." But, of course, it does make a difference. The fact that gender has been rendered invisible in the articulation of urban policy (as well as race, ethnicity and sexuality) means that the masculine ideal forms the basis for institutional structures and practices (Wekerle 2013). For, as Parker (2016a) notes, neoliberal urban policy "continues to valorize masculinized, elite, white subjectivities and concepts like competition and autonomy instead of care and connection. Associated patterns of resource allocation follow." Thus, "gendered" issues like childcare, housing quality and affordability, and poverty among single female heads of households receive too little attention and too few resources; they are marginalized and trivialized, with women as mothers seen as an urban problem rather than part of the solution (Brownill 2000).

This is what Tickell and Peck (1996) referred to as the return of the Manchester Men, in which the neoliberal move to privatize much of urban governance leads to business-oriented growth coalitions. The leading proponents of neoliberalism are men, with the specific goal of maintaining their male privilege (Braedley and Luxton 2010 cited in Wekerle 2013). Women are at a disadvantage in appointments to these groups through patronage (Brownill 2000). But, as Tickell and Peck (1996: 596) note, "Regendering local governance is not just about packing committees with men, it is also about privileging masculinist forms of decision-making and agenda-setting." These groups have undone much of the feminist work of the 1980s in local government. As one of Brownill's (2000: 124) informants in her case study of the redevelopment of the London Docklands put it, "there is something about men in suits that makes me fall silent." In this form of neoliberal governance, the state is patriarchal because women are under-represented, but also because the gendered division of labor is replicated in the state structure, with certain sectors like welfare and education constructed as feminine. Most importantly, the state is patriarchal because it affects the reproduction of gender relations and identity through policies on marital relations, reproduction and abortion, wage discrimination, housing policy, male violence, and so on (Tickell and Peck 1996). One recent example of how this plays out is in the case of the Focus E15 Mothers discussed in Chapter 2. When the mothers first approached their council mayor about their campaign to preserve affordable housing for young mothers, as one mother recalled, he told them, "'I think it's disgusting what you're doing' … And we was completely shocked and at that point we was still quite weak, we were a bit shaken up by him, when we got out of the meeting we were actually in tears, cuddling each other, 'he's our Mayor'" (quoted in Watt 2016: 305). The "Manchester Men" attempt the invisibilization of gender and care issues in the work of the gentrifying city.

The 2008 (and beyond) global financial crisis helped to reinstitute entities that operate like the "Manchester Men" Tickell and Peck (1996) describe. Following the collapse of global financial and real estate markets, many countries created "bad banks" to manage high risk investments that had defaulted. The purpose of these banks was to recoup whatever money possible and protect the rest of the financial system. One such example is that of the National Asset Management Agency (NAMA) in Ireland, which emerged not only as a force to protect financial assets, but a means by which to accomplish gentrification in selected sites by transferring planning power away from local government, increasing the role of private developers, and orchestrating property-led development projects (Byrne 2016). The major focus of these plans has been the Dublin Docklands, a formerly industrial waterfront site (see Figures 3.1 and 3.2). While the original purpose of NAMA was simply to get bad assets off the books, by 2013 it had taken on a place-based focus aimed at redeveloping the Docklands with Grade A office space to attract international investors. The site has office space for Google, Facebook, and Airbnb, among others. Urban geographer Philip Lawton describes the plan as an effort to make Dublin the San Francisco of Ireland, and even of Europe (interview, July 2016).

Urban geographer Michael Byrne (interview, June 2016; see also Byrne 2016) characterizes the Dublin Docklands as a "catastrophic success." It was a missed opportunity, in the wake of the crash, to rethink the relationship between finance and real estate. It was, however, successful in globalizing the

Figure 3.1 Dublin Docklands

Figure 3.2 Dublin Docklands

real estate industry in Dublin (Byrne 2016). Land was sold fast, primarily to venture capital firms in the U.S., with no provisions for sustainable planning or affordable housing in the first wave. While generally deemed successful in terms of property-led development, the role of NAMA in urban development raises troubling questions about process in post-crisis urban planning. As the landowner in the Docklands, NAMA was able to overrule the Dublin County Council, overriding the democratic process. "Fast track planning" decisions in the Strategic Development Zone (SDZ) may not be appealed by citizens in order to provide "planning certainty" for investors (Byrne 2016). In effect, in creating NAMA, the government got rid of its capacity to shape urban development, so that now NAMA is doing public policy (interview with Michael Byrne, June 2016). Long-term residents in the area feel largely marginalized and disconnected from the wealth that surrounds them. Indeed, NAMA gerrymandered its development boundaries so as to avoid having to address issues of social housing, and as a developer, NAMA has no social remit, unlike previous development agencies in the Docklands (Byrne 2016; interview with Philip Lawton, July 2016).

Just as Tickell and Peck (1996) show in Manchester, the process in Dublin was masculinist (interview with Michael Byrne, June 2016). The financial industry that dominates NAMA is almost entirely male. The process of how members are appointed to NAMA is a mystery. The board has seven members (with additional strategic committees as needed). One is a woman. All are drawn from the ranks of the financial sector and political loyalists (Ross 2014). The property-led development model of the Docklands took power away from established community groups which had women in leadership positions (Michael Byrne interview, June 2016). So the work of urban planning and development is re-masculinized, the goal to attract the creative sectors of the economy, now seen as the only, or at least the most important, driver of urban development in this most recent iteration of entrepreneurial urbanism. Meanwhile, cities rely on women's informal labor to replace municipal services that have been cut under the neoliberal regime (Wekerle 2013).

Who is this city for? As Parker (2008) details, much of what is celebrated about the new "creative" economy that has brought gentrification is decidedly masculinist, creating and reaffirming gendered divisions of labor and sexist practices rather than creating any sort of liberatory potential. Richard Florida's (2002) conception of the "creative class" as the driver of urban growth and revitalization tracks neatly with the way in which cities have been accomplishing gentrification, celebrating the work done outside the home in male-dominated industries while eliding the raced, classed, and gendered inequalities in these sectors of the economy and ignoring the social reproduction work necessary for their success. Florida's focus on individuality, meritocracy and diversity serves to reify existing systems of discrimination and create further disadvantage by prioritizing certain workers over others (for a fuller critique of Florida, see Peck 2005).

Parker (2008: 207) describes the hero of the creative class discourse:

He is creative and consumption-oriented. He is fit, finicky, and flexible. He is talented, transcendent, and time-deprived. He can locate himself wherever he pleases, and the city (and now country) that fortuitously snags him is guaranteed a prosperous future.

While this masculinized conception is "softer" than the musclemen of Rotterdam mentioned above, it is no less effective in coding certain people as desirable. As Parker (2008: 208) recognizes, the discourse of the creative class "reifies a particular type of 'hyper' capitalist and individualist worker, whose characteristics, networks, and privileges are particularly exclusive and masculinized." The desirable characteristics of the creative worker are positioned as scarce and valuable, feeding the discourse on urban competition, but also actively devaluing those who don't, or can't, fit this masculinist, "macho" mold (Parker 2008).

While diversity and tolerance are celebrated in the creative city, the vision of diversity is a decidedly narrow one. Florida focuses specifically on the presence of gay households and foreign-born immigrants, with gays the "canaries" that signal a tolerant city. No attention is paid, for example, to the presence of, and outcomes for, African Americans in creative cities. Even Florida (2016) himself has recognized this gap, finding that the creative economy and the creative class both skew white and that there is a correlation between the creative class and income inequality and segregation.

Parker finds that the poverty rates for black and female headed households of all races are high in creative cities. The gentrification that results from an influx of the creative class will lead to the displacement of these households. Thus, "social inclusion for some may entail the social exclusion of others" Parker (2008: 217). The reliance on the market and the dismantling of the welfare state have redistributed income from women to men and required more caring labor from women (Wekerle 2013). Indeed, in dedicating scarce urban resources towards the creative class elite, women, families and minorities in cities disproportionately suffer the consequences (Parker 2008). In the era of "austerity urbanism" (Peck 2012) that has accompanied the rise of the creative class and the installation of gentrification as global urban policy, the retrenchment of the public sector has particularly targeted African-American workers, especially women. And once unemployed, black women were least likely to find private sector employment and most likely to exit the labor force (Lowrey 2016).

Also excluded from the picture is any acknowledgement of the role of social reproduction and care work. The work of the household, parenting, and other care work are not recognized as "creative" (Parker 2008) (more on social reproduction later in this chapter and in Chapter 4). But whenever care work is acknowledged, it is most decidedly gendered. It is no accident that the vast majority of voices associated with personal assistant apps are female: Siri, Cortana, Julie, Clara, Crystal, Jeannie, Cloe, Dawn (named after Don Draper's assistant on "Mad Men"), and Donna (an assistant on the "West

Wing"). As writer Joanne McNeil (2015) argues, these serve to reinforce old stereotypes about what gender is best suited for administrative work and care work in general. There is a maternal edge to all of it (McNeil 2015). But the gendered division of labor behind this new technology, and indeed behind many of the labor sectors that fuel gentrification, remains masculinist. Men are the creators of these apps. Among the 12 combined co-founders of Siri, Julie Desk, X.ai, and Clara Labs, there is only one woman (McNeil 2015). This points to a persistent problem that is the subject of the next section, the wage gap and the continuing gendered division of labor.

The wage gap and the gendered division of labor

Lyons (1996) questioned the degree to which the feminization of high status employment was directly related to gentrification. She found that while women's waged labor was important in enabling the household to gain advantage in the housing market, it was less likely to constitute a high-status professional career. Bondi (1999) in her study of two inner-urban and one suburban neighborhood in Edinburgh found persistent gender inequalities in socioeconomic status in all three neighborhoods, despite small gender gaps in educational qualifications. The glass ceiling persists. The Economist (2014; see also Grossman 2014) constructed a "glass ceiling index" to track how likely women were to be treated equally at work. This includes data on education, labor force participation, pay, child care costs, maternity rights, representation in senior jobs, and applications to business school. Even those countries that score highest on the index, Norway, Sweden, and Finland, score just below 80 of a possible 100. The OECD average is below 60. And those ranked lowest on the list (of 27 countries), South Korea and Japan, score at 20 or below. In these nations, pay gaps are large (37 percent in South Korea), there are few female senior managers, and simply too few women in the labor force. The simple presence of women in the labor force is not enough to undermine patriarchy, especially as the increase in women's participation in the labor force came at a time when work was also far more casualized and insecure, a trend that has only increased with time (though, as McDowell (2014) reminds us, insecurity has long been typical of employment patterns for women).

Jobs too have gentrified (Mock 2015a). One study found that high-skilled workers have moved down the occupational ladder and have begun to per-form jobs traditionally performed by lower-skilled workers, thereby pushing lower-skilled workers farther down the occupational ladder, so that, "having a B.A. is less about obtaining access to high paying managerial and technology jobs and more about beating out less educated workers for the barista or cle-rical job" (quoted in Cassidy 2015). Employers from investment banks, law firms, and other segments of the economy associated with the new middle class and gentrification, in which employment requires a college degree, prefer to recruit from upper tier colleges but don't pay much attention to things like majors and grades. "It was not the content of education that elite employers valued but rather its prestige" (quoted in Cassidy 2015). This job gentrification further disenfranchises those with the least ability to compete. In the U.S., only 14 percent of African-American job applicants are called back by employers, with similar statistics for African-American women and Latinas (Mock 2015a).

Warren and Tyagi (2003) argue that when mothers entered the workforce, families gave up the economic value of an extra adult to pitch in to financially support the family in times of crisis. The stay-at-home mother was a financial safety net, available to enter the labor force when extra income was needed. Women entered the labor force as the state was retrenching, so rather than saving the second income, families spent the extra income on the search for homes in good quality school districts in neighborhoods seen as safe, applying inflationary pressure to real estate prices. Thus, double-income households actually have less discretionary income than single earner households a generation ago (Warren and Tyagi 2003). This pushed single income, female headed households farther down the economic ladder.

Compounding this problem is the persistence of the wage gap, which is inseparable from the persistence of the gendered division of labor. The fact remains that work done by women simply is not valued as highly as that done by men, with employers placing lower value on the work done by women (Miller 2016). Much of the wage gap can be explained by the difference between the occupations and industries in which men and women work (Miller 2016; Blau and Kahn 2016). So, for example, the median earnings for information technology managers, who are mostly men, are 27 percent higher than human resources managers, who are mostly women. At the other end of the wage spectrum, janitors, a typically male occupation, earn 22 percent more than the female dominated jobs of maid and housecleaner (Miller 2016). Of the 30 highest paying jobs, occupations like chief executive, architect and computer engineer, 26 are male dominated. Of the lowest paying jobs, like housekeeper and child care worker, 23 are female-dominated (Miller 2016; Liner 2016). The gap is not simply the result of gendered choices in the labor market. When women start moving into occupations, those jobs begin paying less, even after controlling for education, experience, skills, race and geography (Miller 2016; Levanon et al. 2009). As economist Paula England notes, "It's not that women are always picking lesser things in terms of skill and importance ... It's just that employers are deciding to pay it less" (quoted in Miller 2016).

Economist Claudia Goldin (Goldin and Dubner 2016) argues that the wage gap results not primarily from overt discrimination, though as we have seen that certainly does exist, but from the high cost of temporal flexibility in many occupations. The gap exists because hours of work in many occupations are worth more when given at particular moments and when the hours are more continuous (Goldin 2014). Because the burden of care so often falls on women, and the assumption is that it will fall on women even if they are not yet engaged in care work, those jobs that offer some measure of flexibility pay less than those with less flexible hours, even if employees end up working the same number of hours. Flexibility has become simply a way to manage long hours, rather than achieve any real work-life balance (Yerkes et al. 2010).

Women who need this flexibility to do care work pay a price, literally. According to Warren and Tyagi (2003), having a child is the single greatest predictor that a woman will end up in financial collapse. The "mommy tax" or "motherhood penalty" is the price women pay in the labor force for engaging in care work, and it is prevalent across national boundaries. The penalty remains even after adjustments for foregone work experience, education and reduced work hours (Budig et al. 2012), a situation that is exacerbated for lone parents and low-income families (Yerkes et al. 2010).

Cultural expectations remain a powerful force. In countries where cultural attitudes support the male breadwinner/ female caregiver model, even positive public policies like parental leaves and public childcare have a less positive, or even negative relationship, to maternal earnings (Budig et al. 2012). Budig et al. (2012) find that leave length impacts employers' perceptions of mothers. Moderate leaves reduce pay gaps, but no or short leaves increase pay gaps, as women may chose to leave paid employment altogether, while long leaves are linked with decreased employment continuity and earnings, as employers may question a woman's commitment to work. Meanwhile, high levels of childcare positively affect women's labor force participation, but there is great variability in how well childcare hours map onto normal working hours.

Even in the Netherlands, often cited as a "best practices" model, there is a significant disconnect between stated government policy and women's lived experiences. In their comparative study of the relationship between national policy and lived experience in the UK and the Netherlands, Yerkes et al. (2010) find that policy in both countries continues to frame women as the primary caregivers and is largely ineffective in meeting the needs of working parents, doing nothing to encourage men to take up caring work. It is taken as a given that women will integrate care with work and that women's work is therefore more open to disruption than men's. They find that this gendered division of labor was even more extreme if women were divorced or separated. Indeed, they argue, "the fact that women are now economically active but remain primary caregivers within the domestic sphere questions the proposition that the labor market and the gendered division of labor have been transformed at all" (Yerkes et al. 2010: 416).

Boterman and Bridge (2015), in their comparative study of parenting as social relation in London and Amsterdam, come to a similar conclusion, finding that gender roles, especially in relation to child care, are remarkably uniform across their neighborhoods of study, thereby partly contradicting the notion that gentrifiers would have more egalitarian gender practices. Across all classes, they found that paid work was strongly gendered, particularly among parents with young children. They do find a more equitable division of labor in Amsterdam, boosted by national policy that is supportive of both part-time work and the provision of childcare. Particularly in the public sector, working four days a week is not an impediment to achieving top-ranked positions. But, in highly competitive sectors like financial services and major law firms, part-time work remains less accepted, thereby limiting parents' ability to achieve a more equitable division of labor (Boterman and Bridge 2015).

Goldin (2014) found that the work environments that require more interactions or have more time pressure are those with larger gender earning gaps. So, for example, any time off for MBAs is heavily penalized. Yet, it is women in these jobs with the highest penalty to time out who took the most time out, largely because their jobs did not enable shorter or more flexible schedules (Goldin 2014).

Because certain occupations impose heavy penalties on employees who want fewer hours and more flexible employment, the resulting lower remuneration can result in shifts to an entirely different occupation, or to a different position within an occupational hierarchy, or to being out of the labor force altogether (Goldin 2014). In the UK, where national policy is less supportive of working mothers, mothers are less likely to work than in the Netherlands (Boterman and Bridge 2015). In the U.S., Anne-Marie Slaughter (2015) finds that we are losing women from the work force. American women are now outperforming men in school and entering the labor force at higher salaries. The wage gap is not evident at this stage, as women have increased their productivity enhancing characteristics and "look" more like men (Goldin 2014). But while women make up more than 50 percent of entry level positions, they are just 10 to 20 percent of senior positions where it becomes clear that any sort of work-life balance cannot be sustained (Slaughter 2015). They are not leaving their jobs but rather being pushed out by workplaces that refuse to allow accommodations for family life.

This is not a women's problem; it is a work problem (Slaughter 2015), but the gendered division of labor is such that this burden still falls disproportionately on women. What is required is a change to the nature of work. Greater flexibility across occupations is required; just changing the gender mix of occupations will not do the trick (Goldin 2014). Goldin proposes greater flexibility across employment sectors as easier, cheaper, and more feasible than getting men to change their behavior by performing an equal share of domestic duties (Goldin and Dubner 2016).

While this phenomenon is not exclusive to gentrified neighborhoods, these were the kinds of jobs that created the demand for gentrification. Yet, the gendered wage gap declined much more slowly at the top of the wage distribution than at the middle or the bottom and by 2010 was noticeably higher at the top (Blau and Kahn 2016). Gentrification was supposed to be a potential strategy for finding this work-life balance, allowing the new middle class to live closer to work and services like childcare, schools, and shopping that would enable a better quality of life. But the increasing costs of housing and quality childcare combined with increased economic insecurity even at the higher levels of the wage ladder have fostered an increasingly competitive work culture that demands more while providing very little support. Care work is invisibilized, commodified, and undervalued.

Commodifying care

As Markusen (1981: 41) argued, "the value of women's labor time in household production can only be accurately assessed if full value is accorded women's work in the wage-labor market." The stubborn persistence of the gender wage gap demonstrates that this labor is still not accorded full value, and therefore the labor women do in household reproduction remains undervalued as well, even as what was once unpaid labor becomes more fully part of the waged labor market. This social reproduction work is systematically undervalued, gendered, and often raced. Any occupation that involves caregiving pays less, even after controlling for the disproportionate share of female workers (Miller 2016; England and Folbre 1999). Data from the U.S. highlights the inequity. Ninety-five percent of child care workers are female; annual median earnings are $22,984. Eighty-four percent of housekeepers are female; annual median earnings are $20,852. Eighty-eight percent of nursing and home health aides are female; they earn a median income of $24,544 (Liner 2016).

As McDowell (2014) argues, one of the most significant trends correlated with women's rising participation in the labor force has been the transfer of domestic labors into the market where they are commodified; it is work still largely undertaken by women and remains undervalued and underpaid (see also Brown 2011). Contrary to mainstream economic theories about the geography of globalization and the shift towards affective, creative, and immaterial labors beyond the workplace, caring labor is embodied, hands-on labor in which productivity gains are all but impossible (McDowell 2014). This care work, undertaken by women and therefore low paid, reflects the "natural" attributes of femininity, rather than any skills acquired through training (McDowell 2014, 2003). Workers engaged in this labor come to be seen "like family" and are thus often asked to do more for less (Brown 2011; Pratt 2004).

The success of liberal feminism, as reflected in the greater access for some women to high-paying professional careers, contributed to the increasing importance of class differences (which are themselves raced) among women and the polarization of households in gentrifying neighborhoods between dual income professional households and those with only one low or modest income (Rose 2010). As economic disparity in urban centers increases, the rich hire the poor to do private domestic work, thus supporting further economic disparity (Brown 2011). As one employer, who was also an organizer for domestic workers, cited in Brown's (2011: 146) study of West Indian nannies in Brooklyn commented, "I was shocked at the lack of consciousness of women employers towards women employees … When feminists go to the workplace, they have an idea of how they want or expect to be treated, but then go home and exploit the women who work for them."

While working women with children or dependent relatives face the common problem of "time-poverty," upper income women have vastly more resources to cope with the situation by living closer to work in gentrified neighborhoods and substituting paid services for previously unpaid domestic and caring labor (Rose 2010). They are able to purchase the labors of working class women to replace their own domestic labor in the form of cleaners, nannies and child minders, therapy of various kinds, and the growth of options for eating outside the home (McDowell 2014). Upper class families benefit from the undervaluing of women's work, as it gives them ample opportunity to hire poor women of color who are often first generation immigrants (Brown 2011). This caring division of labor is global and thus the ability to purchase replacements for domestic love and service has expanded exponentially (McDowell 2014; see also Nast 2016).

For many middle and upper class women, any attempt at work-life balance comes from the availability of cheap domestic labor which is overwhelmingly female and often immigrant, a new servant class (McDowell 2014; Brown 2011), created through a process of "deskilling through immigration" as immigrant women are ghettoized within marginal caring occupations (Pratt 2004). The co-construction of employment characteristics and types of femininity has created migrant women as particularly appropriate employees in the provision of commodified caring, with millions of women moving from south to north to care for the children and elderly relatives of wage earners (McDowell 2014). "[T]he intersection of class, gender, ethnicity, and nationality combine to construct a hierarchy of difference in the labor market in which caring labors are not only poorly rewarded but are increasingly associated with a division of labor in which women migrants are the least eligible" (McDowell 2014: 8).

The racialized hierarchy within this division of labor, one that intersects with gender, nationality and class influences how care work is both "performed" and consumed (McDowell 2014). In Pratt's (2004: 55) research, European nannies were seen as more intellectual while Filipinas were seen as less demanding, "these women [Slovakian nannies] are intellects, and you're not going to have six intellects living in a house at the weekend, sleeping on the floor. Whereas with Filipino nannies, they love to roll all over each other all weekend!" This refers to the common practice of sharing rental accommodations on the weekend. Similarly, one employer fondly remembered a Filipina domestic worker who "survived on toast and air," (quoted in Pratt 2004: 55) while one employment agent announced, "you want someone to lick your home clean: Filipino girl. Go for that" (52). Tamara Mose Brown's (2011) study of West Indian nannies in Brooklyn found that this ethnic and racial "grading" functioned to extract more labor from child care workers. Recognizing their position at the bottom of the labor market, many of the women in Brown's study went out of their way to accommodate employer demands, far beyond their official responsibilities.

As McDowell (2014: 8) notes, "These performances draw on cultural repertoires but are not voluntaristic." They are produced within regulatory frameworks and must conform to managers' and clients' expectations. These gendered and racialized norms and expectations affect men as well. Care work is considered appropriate within the construct of certain racialized masculinities, leading Pratt (2004: 81) to argue that racialized men can be more profoundly feminized than a middle-class white woman.

Much of the flexibility and creativity we celebrate in the postindustrial economy would not be possible without this largely invisibilized labor. But too often, as in the creative class discourse, there is no recognition that an upper class household's participation in the creative economy is predicated on the reproductive activities of racialized women. No attention is paid to the social services and institutions that facilitate social reproduction (Parker 2008). Indeed, the flexible work day of the creative class has made the work of child care more difficult, not less, requiring child care workers to be similarly flexible, with unpredictable hours and precarious pay, as reflected in one worker's summation of her work situation, "They're using me like a donkey" (quoted in Brown 2011: 50).

The flexible workplace celebrated by the creative class also makes child care more difficult for providers, with many of them forced to spend the day outside with the children so that their creative class employers may work from home (Brown 2011). This leaves nannies traveling between parks, libraries, children's classes and other public space in all kinds of weather. It also opens up these workers to intense surveillance, especially when the caregivers are racialized others (Brown 2011). Technology exacerbates this, with employees required to be always available by cell phone, subject to "remote mothering," monitored by nanny cams, parenting blogs, stroller license plates, and websites like isawyournanny.com (Brown 2011).

So we see precarity at multiple levels of the job market in ways that reproduce and reinforce the need for low paid, "flexible" labor. These roles are gendered, raced and classed in different ways for different people and in different sectors of the economy. In the current neoliberal era, few feel safe from this precarity; it has extended to many areas of the economy previously considered secure. This may provide new openings for resistance. For McDowell (2014: 19),

> Taking seriously the intersection of class, gender, ethnicity, nationality, and skin color also challenges the traditional notions of class and of class struggle, and may lead to a richer understanding of inequality as well as the basis for building stronger alliances across social divisions.

One such example of the potential to build alliances across traditional class lines is the Fight for 15 movement in the United States. The movement started with a focus on achieving a living wage of $15 an hour for fast food workers but has expanded to a larger movement looking at income inequality across the economy, drawing support and making alliances with other precarious workers like child care workers and contingent faculty at American universities, attempting to build a "moral solidarity" across sectors of the economy for workers experiencing the same insecurity (Flaherty 2015).

Gentrification relies on the underpaid labor of the restaurant workers, delivery people, child care workers, dry cleaners, housekeepers, and others who make it possible for those in high end service sector jobs to work the long hours required to afford the gentrified city. But this balance is precarious for all those involved in it. Seeing the common ways in which labor is exploited, and gendered, across the economy provides the potential to do work differently, to value work differently. With the majority of American millennials, for example, identifying as working class (Malik et al. 2016), there is perhaps a growing recognition that the way things work now is simply not working.

Conclusion

The gendered division of labor and the undervaluing of care work continues to create sexist cities, a problem that gentrification, once thought of as a potential solution, has helped to reinforce, albeit in different ways from those of the industrial city. The feminization of the urban economy and landscape, with a focus on consumption, has been a tool by which to displace the "masculine" industrial, the sweaty, dirty work that was often unionized and was more likely to provide a living wage than the postindustrial "feminine" service economy that has taken its place. A feminine face is often put onto the work of redevelopment even as those making the decisions and benefitting from the process are overwhelmingly male, with masculinist agendas that privilege profit over care work and service provision.

This division of labor in urban planning for gentrification reflects and reinforces the gendered division of labor, with women's work being undervalued by virtue of the fact that it is women who do it. The inadequate provision of child care and other care infrastructure further disadvantages women and others with care responsibilities, with the flexibility that is so often hailed as the hallmark of the new economy actually serving to systematically underpay women. This inequality travels down the economic ladder, with upper income families relying on the low wage work of feminized immigrant labor to allow them to work outside the home. At every stage, the increasing flexibility and precarity of work requires longer, unpredictable hours with uncertain and inadequate pay. The delicate balance is predicated upon social reproduction work that goes unseen and undervalued, and that itself often serves to reinforce the gendered division of labor. This social reproduction work is the topic of the next chapter.

4   Social reproduction

The patriarchal assumptions that underlie urban design continue under gentrifi-cation. As Boterman and Bridge (2015: 249) recognize, "Urban space is a situ-ating framework and an active process in trajectories of social reproduction." The continued separation of work from home, the concentration of services in downtowns, the plan and schedules of public transportation, and the lack of facilities for children in many neighborhoods all assume a particular trajectory of the life course and organization of space based on the past experience of women confined to the home and urban areas as not ideal for children. Even as these demographic facts change on the ground, urban structures remain the same, often exaggerated by gentrification, for example the intensification of high rise development discussed in Chapter 2. Rather than facilitate the dual role for women and others in the position of caregiver, the organization of urban space and urban policy under gentrification serves to increase the challenges of social reproduction.

Space authorizes some performances and hides others (Aitken 1998). Under neoliberalism, the grand and the spectacular are more in focus than the small and the ordinary (Cele 2015). This is especially true in the realm of social reproduction, which, following Katz (2001: 709), I understand as "the material social practices through which people reproduce themselves on a daily and generational basis and through which the social relations and material bases of capitalism are renewed." This work is rendered nearly invi-sible under neoliberalism. Care work and other forms of social reproduction become visible only when they interact with the market, and the increasing dependence on the market under neoliberalism is not an even-handed policy (Betancur and Smith 2016). This chapter address the ways in which the (re) making of urban space through gentrification, the intensification of mother-hood, the privatization of education, and the privatization of urban govern-ance have served to make social reproduction more difficult, more masculinist, and less democratic. I will focus here on the organization of urban space and its effects on the rising costs of parenting, the gentrification of schools, and community organizing, all aspects of urban life that are highly gendered and rely on longstanding masculinist assumptions about the gen-dered division of labor. But, I argue, the work of social reproduction also allows the opportunity for alliances across gender, race, class, sexuality, and ability to reimagine who the city is for and how urban space should be organized.

Gentrification of parenting and childhood

The early literature on the role of women in gentrification saw gentrification as a spatial solution to the problem of work-life balance. Women, especially those with care responsibilities, would be attracted to central city locations in order to be close to their jobs and manage their domestic work all while having access to the amenities of the city. But the urban spaces that gentrification has adapted and created tend not to recognize the dual role of work and care. The narrative of urban living for the affluent tends to minimize, or ignore altogether, the role of care and family in urban design. As Stamp (1980: 189) asserted, "the neighborhood of the American city generally does not meet basic human needs for support and belonging. It was not designed to do so." England (1991) has argued that urban planners have failed to act on the predictions that women's move into the labor force would require more flexibility in urban design, so that women and other caregivers are forced to adapt to the existing spatial structure rather than the other way around. This is not just a historical legacy. As discussed in Chapter 2, new housing being created in gentrified neighborhoods in cities around the world continue to assume a particular life trajectory that privileges the young and single without children or empty nesters as urban residents while ignoring the needs of families, and indeed, ignoring the role of gender in urban planning altogether (Fincher 2004; Wekerle 2013). As Cele (2015: 241) argues in her study of Stockholm, the failure to include play spaces, preschools and other necessary amenities is "a deliberate social and material exclusion of the young" and, by extension, those who care for them. The neoliberal assumption that the city is engaged in a global competition, and that this necessitates specific kinds of development, affects conceptions of citizenship, community and everyday life which inevitably shape children's access to the city (Cele 2015)

These assumptions were made clear in a debate in Toronto over a proposal requiring a developer to include three-bedroom apartments in a new development, an explicit attempt to provide more family-friendly housing in a city with a booming population. The then-deputy mayor, Doug Holyday, fought the proposal, asking, "How much social engineering are we going to put into this city and at what cost?" His opposition was based on the fact that he did not see the central city as an appropriate place to raise children. "[T]here are healthier places to raise children," he argued, "I can just see it now, 'Where's little Jenny? Well, she's downstairs playing in the traffic on her way to the park'" (quoted in Church 2012). Similar battles have emerged around the issue of strollers (Mallick 2013) with some municipalities requiring that strollers be folded to board public transit or not allowed when buses are crowded or during rush hour. While strollers certainly do take up space, this is because they are holding actual people, though children often are not recognized as fully deserving of the rights of citizenship in these debates. Thus, central cities can be inhospitable to people with children and children themselves, with too few amenities like public parks and accessibility challenges in many buildings (Harrison 2012).

This has been my own experience as the mother of twins in Chicago. I wrote an op-ed about the gendered effects of the city's plowing policies during the city's notorious winters (Curran 2014) commenting that the priority of plowing arterials and other major streets first left the streets on which many schools and childcare centers are located unplowed, and therefore difficult to navigate for caregivers, who remain, overwhelmingly, women. The assumptions behind these sorts of policies, as with the example of strollers on public transit above, prioritize those getting to and from work without other responsibilities, and ignore the other aspects of life.

These oversights are difficult to navigate even for those most resourced, but we see the cost, both financial and personal, of raising children in cities increasing, often as a result of gentrification. Certain families have become catalysts for gentrification (van den Berg 2013), with young urban professional parents (YUPPs) choosing urban parenting as the way to manage their time-spatial budgets to be near many facilities, including work (Karsten 2014). For many families, though, following the Great Recession, the choice to remain in the city may not be much of a choice, with families stuck in housing they cannot sell and/or unable to access financing to buy the traditional house in the suburbs.

The rising costs of housing resulting from gentrification put the cost of adequate housing out of reach for many families while the cost of child care in countries where it is not subsidized is not sustainable, often forcing women out of the labor market, as their wages are lower than the costs of childcare (Quart 2013). A 2014 poll found that fully 54 percent of urban parents in the U.S. said they just meet or don't have enough to cover basic expenses (compared to 38 percent of non-urban parents) (Marshall 2014). The Economic Policy Institute estimates that a two-parent, two-child household in New York City needs $93,502 for "a secure yet modest living standard" (Badger 2013). In liberal welfare states, policy related to family and child care rests on the "primacy of the market and the privacy of the family" (Sainsbury (1999) quoted in Warner and Prentice 2012: 197). Care is an individual responsibility; provision of that care is structured by the market.

The problem with this neoliberal model is that just as mothers are being more fully drawn into the capitalist labor market, ideological constructions of intensive mothering increase their workload at home (Holloway and Pimlott-Wilson 2016). Neoliberalism relies on ideologies of intensive mothering to legitimate the low-cost reproduction of future citizen-workers by women at the same time that labor market insecurity has been heightened for all workers, and the lines between work and home have been blurred (Holloway and Pimlott-Wilson 2016). As in all class-riven contexts, these idealized notions of home and mothering are hegemonically produced in the interests of those who can afford or benefit from them (Nast 2002).

The response of gentrifier parents to these challenges has resulted in the privatization and intensification of parenting that makes child rearing more expensive and labor intensive for everyone. This intensive parenting requires "lavishing copious amounts of time, energy, and material resources on the child" in "*child-centred, expert-guided, emotionally absorbing, labor-intensive,* and *financially expensive*" childrearing (Hays 1996) quoted in Holloway and Pimlott-Wilson 2016, emphasis in original). This ideology is shared across class backgrounds even as it unachievable for many (Holloway and Pimlott-Wilson 2016).

Jezer-Morton (2015) argues that this helicopter parenting is not a set of values; it is a nervous condition born of the rising cost of being a middle or upper class parent in the gentrified city. Economic pressure from high rents, the high cost of child care, and the competition for quality education in a precarious economy lead to fierce pressure to succeed in the work of raising children. As Katz (2008) has argued, the anxiety over the economic future is channeled into the commodification of children. "Children are the 'new investments' in middle class families" (quoted in Jezer-Morton 2015). Brown (2011: 71) recounts her own experience of parenting in a gentrified neighborhood in Brooklyn, reasoning that if you love your child enough to pay for lessons, other neighborhood residents consider you a "good mother." It functions as a demonstration of social capital. This intensive parenting is part of a broader process wherein income inequality, the root cause of helicopter parenting, is coming to be considered part of the natural social order of life (Jezer-Morton 2015). This phenomenon is global (see for example Faircloth et al. 2013; van den Berg 2013; Boterman and Bridge 2015).

Parenting practice remains highly gendered, often despite many gentrifiers' best intentions. Bondi (1991a, 1999) has long argued that gentrification is a process of both gender and class constitution. Her (1999) study in Edinburgh found that raising children challenged the aspirations to gender equality between partners with no evidence of variations in gendered practice between the suburban and gentrified neighborhoods, thus undermining the argument that gentrification would somehow challenge the patriarchal division of labor. As novelist Rachel Cusk (2001:5) recounts in her experience of motherhood, a day spent at home is on the other side of the world from the office. "From that irreconcilable beginning, it seemed to me that some kind of slide into deeper patriarchy was inevitable." Instead of a reworking of gender roles, we see the increasingly visible commodification of mothering played out on the streets of urban neighborhoods, with phenomena such as the "yummy mummy" in Britain providing the demand for baby clothes shops, boutique toy stores, and cafés in which mothers can socialize (Boterman and Bridge 2015). Urban space is being feminized, but not in any way that challenges conventional gender norms. It transforms motherhood into a consumption strategy, countering the traditional isolation and boredom of child rearing with expensive strategies to privatize public space.

The consumption practices of middle class parents have made parenting much more visible in certain urban spaces, with public parenting widely on display in neighborhoods that have been gentrified by families (Karsten 2014). Middle class families demand the resources to repave streets, replant parks, and build new playgrounds in their neighborhoods. Developments like these are highlighted in local media, with one article about the construction of an urban beach in Amsterdam titled, "East is becoming the Brooklyn of the city," reproducing a global image of the city as a place for (certain) families to live (Karsten 2014).

This investment in some urban families comes at the expense of under-served areas of the city. While the urban neighborhood has long served as a site of social support for mothers, the increasing colonization of space in gentrified areas may serve to further disenfranchise working class mothers and other caregivers who cannot afford to participate in the café and baby class culture. The changes to the commercial street visible in gentrifying neighborhoods make clear who belongs and who doesn't, a project that is classed, raced, gendered, aged and abled, with marginalized groups requiring a wider range of commercial personal services, like laundry and internet access, than middle class households (Rankin and McLean 2015). Everyday commercial spaces function as key sites of social reproduction (McLean et al. 2015). Commercial displacement is just one of many strategies by which working class families are priced out and made to feel they do not belong in their own neighborhoods.

The intensive parenting made so visible in commercial and public space by gentrifier parents raises the bar for what is considered adequate parenting in a competition that many working class parents, single parents and others cannot hope to win. As Jill Lepore (2016: 50) has commented in her history of child welfare policies, "the more the children of the better-off were cherished, and pampered, the worse the treatment of the children of the poor appeared to be." In gentrifying neighborhoods, concern about child welfare becomes a strategy through which gentrifiers object to the different class and racial backgrounds of their neighbors. Martin (2008: 343) found in her study of gentrifying neighborhoods in Atlanta that class and race differences between gentrifiers and long-term residents were "effectively cloaked as differences in child rearing, and categorized as either good or bad." As van den Berg (2013) notes in her study of Rotterdam's attempt to attract young middle class families with children at the expense of immigrant and working class families, while the children of the middle class are imagined as the solution to urban problems, working class mothers and children are seen as the cause of urban problems. Under neoliberalism, the responsibility for this care and development lies solely with the parent. This allows the state to lay the blame for unequal child outcomes at the door of the family rather than an inegalitarian society (Holloway and Pimlott-Wilson 2016).

This privatization of responsibility and stigmatization of working class care becomes institutionalized in public welfare institutions, and most especially in schools. The gentrification and abandonment of public education is a striking example of how the gentrification of social reproduction reifies class, racial, and spatial divisions.

Gentrification and schools

Schools are often the last frontier for gentrification in a neighborhood (Hankins 2007), as the young, single professionals who are seen to drive the process settle down, have children, and then strategize on how best to maximize the life chances for their children. The focus on schools comes after the new housing, the coffee shops and restaurants, the improved public transportation, sanitation, and policing. Gentrifying schools has become an essential strategy to keep middle class families in the city. In their work on school choice in London, Butler and Robson (2003) argue that education markets rival those in housing and employment in determining the nature and extent of gentrification, with good schools causing a rise in house prices between 15–19 percent in their catchment area. Similar findings come from Washington, DC, where homes in districts with high rated schools are a third more expensive than those in districts with poor schools, with one city council member recognizing "As soon as you turn the schools around in any neighborhood, that neighborhood is gentrified …It's very powerful" (quoted in DePillis 2014).

The geography of school improvement varies. Gentrification does not automatically make neighborhood schools better. Indeed, gentrification may make public schools worse (Keels et al. 2013). The middle class adopts a strategy in which the whole metropolitan area serves as a single market in which they identify the best opportunities, much as they do with the job market; if their local schools aren't "good," they use their considerable resources to get their children into better schools elsewhere (Butler and Robson 2003; see also DeSena 2006). One result is the segregation by class, and by extension race and ethnicity, in local public schools, and the increasing demonization of those schools (DeSena 2006; Hamnett and Butler 2011). It also may have the effect of hollowing out existing neighborhood schools. Most schools receive funding per child; if enrollment drops, the school has fewer resources. At the same time, those upper-income parents in the public school system, often in charters, demand greater resources, drawing resources away from those who need them most (Hannah-Jones 2015).

Negotiating the system of education is one way women gentrifiers express their gender and relative social class, as it is still overwhelmingly women who not only are the primary caregivers of children but also research, analyze, and select schools for children. In this, they are "doing" social class and social status (DeSena 2006: 254). Their social identity is reproduced by sending their children to more upscale schools, thereby also reproducing social stratification. Their children are taught and learn that they are privileged compared to their working class neighbors (DeSena 2006).

While many long-term residents do not dispute the problems with public education, they see blanket statements about the quality of schools as reflecting poorly on them, their children, and their decision as parents to send their children to these schools (Martin 2008). In this context, education "reform" reinforces existing prejudices and inequalities. This discourse is racially coded, designed to pathologize and discipline minority communities and their institutions (Lipman 2009; see also Boterman 2013; Gulson 2007; Huse 2014). Neighborhood school quality is understood in the context of what Gulson (2007) calls the "absent presence" of the white middle class, where both school improvement and neighborhood revitalization require middle class, non-Aborignal subjects (in the Australian context, but the pattern is repeated in other racialized contexts). Even in a city as diverse and politically liberal as Amsterdam, for example, the ethnic composition of schools is used as a marker of academic quality (Boterman 2013). Parental choice in schooling is the means by which middle class parents have maintained their privilege in terms of access to high quality education while circumventing the perceived disadvantages of diversity (Butler and Robson 2003; Boterman 2013; Gulson 2007). While the move in education policy towards school choice may create new opportunities for a small number of students, the vast majority, especially students of color, attend schools likely to prepare them primarily for low-wage jobs (Lipman 2002) thereby perpetuating the vicious cycle of poor educational quality leading to poverty and poverty leading to poor school outcomes (see for example Stand Up! Chicago 2012). The similarities of strategies across cities indicate that the urban middle classes have, to some extent, similar white identities and constitute a global class (Boterman 2013).

The stigma around existing public schools is so powerful that many cities have started creating parallel education systems to remake (and undo) existing public education. The "school failure industry" (Kotlowitz 2015) is big business. Schools must be branded as failures in order to be remade (Lipman 2009; 2012). "School choice" then becomes the mantra for cities looking to attract white and middle class residents (Hannah-Jones 2015) with the rise of charter schools leading to the creation of alternative community institutions that draw resources away from existing local schools with the explicit goal of keeping the middle class in the area (Hankins 2007) or attracting the middle class to a neighborhood. The existence of a charter school becomes a selling feature in real estate listings (Hankins 2007).

Perhaps no city so exemplifies this neoliberal shift in education policy and its negative consequences as Chicago, the city education researcher Pauline Lipman once described as "the forefront of the rearguard" (cited in Curran and Breitbach 2010: 395). Under a policy called Renaissance 2010, Chicago Public Schools (CPS) planned to create 100 new schools, mostly charters, while closing underperforming schools. The effort was funded, in part, by the Civic Committee of the Commercial Club of Chicago, since the idea for Renaissance 2010 came from the Commercial Club of Chicago, an organization of the city's most powerful corporate and financial leaders (Lipman

2009). Renaissance 2010 is part of a larger trend revealing the ability of corporations to shape the city and schools in their own interest (Lipman and Hursh 2007).

One of the key strategies of school reform has been the demonization of teachers and the attempt to undermine teachers' unions, whose members tend to be overwhelmingly female. In Chicago, where the mayor has negotiated with the police and firefighters' unions but declared war on the Chicago Teachers' Union (CTU), both male and female union members express the opinion that the mayor feels safe taking on the CTU because it is dominated by women (Joravsky 2015). The union representative at my children's school told me that as the union considered a strike in 2016, delegates joked about paying men to stand on the picket lines with them to create a sense of legitimacy. The new charter schools that have been opened with resources taken from existing neighborhood schools tend to be non-union, with lower pay, longer hours, no tenure, and no seniority (though, since 2009, one-fifth of Chicago charter schools have unionized, a trend now gaining steam in other cities as well (Kelleher 2015)).

This privatization of schools, and indeed of urban policy itself, furthers the neoliberal agenda while accomplishing gentrification and displacement. Chicago lost 200,000 people from 2000 to 2010, 90 percent of whom were African-American, the majority of them children (130,000), suggesting that this population is expendable to the life of the city (Betancur and Smith 2016). Rather than doing anything concrete to eradicate poverty or improve educational opportunity, Lipman (2009: 216) argues, these policies instead "demonize, displace, and disperse low-income people." They also increase the everyday oppression of women, since it is they who are primarily the ones who negotiate school placement as well as school transportation and meetings (see Curran and Breitbach 2010). By 2010, while CPS had created 92 Renaissance schools, just 16 of the schools were performing at or above state averages and 3 out of 4 CPS students still attended low performing schools (Clarke 2015). Despite this track record, in 2013 Chicago closed 50 schools, overwhelmingly in African-American neighborhoods of the city, the largest school closure in American history, while still approving new charters. Eighty-eight percent of students affected were African-American (De la Torre et al. 2015). While the schools that were closed had experienced declining enrollments, the communities in which they were located still had a higher proportion of children than the city average and that in higher income white communities (Betancur and Smith 2016).

This turbulence surrounding local schools mirrors that of the housing market for poor and working class Chicagoans. Without a stable place to live, it is impossible to fully participate in school and other community institutions. Gentrification exacerbates this instability both in the housing market (as discussed in Chapter 2) and in education policy. In Chicago, school policy has been explicitly tied to the restructuring of public housing and real estate development, with new mixed income schools aimed at areas with newly redeveloped mixed income communities on the sites of former public housing projects (Lipman 2009). Closing schools in low-income communities and opening new schools is part of the "rebranding" of these communities (Lipman 2009). One Chicago alderman was explicit about this process. Asked what he would do to improve a local school, he proposed this solution, "Quite frankly, what I would like to do is make Manniere 'Franklin II,'" Burnett said. "Franklin is right down the street. Franklin has a long waiting list to get in there ... Make it Franklin II, and I guarantee you we'll fill the school up" (quoted in Cox 2016).

New names obscure the more violent process behind school reform. The new schools frequently have exclusionary selection processes, with complicated applications and informal selection mechanisms that disadvantage less resourced families (Lipman 2009). The school application process in Chicago can be so complicated and fraught that there is an entire cottage industry of consultants who navigate the process for those parents who can afford to pay for their services. New charters have access to resources that traditional public schools do not and are therefore able to market themselves in ways that community schools have neither the resources nor expertise to access, for example, in their flexibility and experience with fundraising and partnerships with corporate donors.

I have seen how the interplay of gentrification and school policy plays out in the neighborhood of Pilsen in Chicago, where I have done anti-gentrification work since 2004 (as discussed in Chapter 2; see Curran and Hague 2006). Whittier School is a neighborhood school, a dual language English and Spanish academy that is 95 percent low-income, 98 percent Hispanic, and 70 percent English language learners (Curran and Hague 2013). It has been continually threatened with closure because of under-enrollment. But this under-enrollment is not simply a reflection of parents choosing not to send their students to a school that has been designated underperforming. Whittier has smaller classes because of the large number of English language learners. Whittier is under pressure both from competition from a new neighboring charter school as well as from increasing housing prices felt by the school's families. While tracking and explaining gentrification-related displacement is notoriously difficult (see for example Atkinson 2000), one potential methodology we employed in Pilsen was tracking where students who left the neighborhood school went. Three hundred and fifty-seven students transferred from Whittier School from June 2006 to June 2010. While some of these students did transfer to the neighboring charter school or public fine arts school that has a regional gifted program, all but 50 of these students moved out of Pilsen, with significant movement to neighborhoods farther south and west of Pilsen with more affordable housing that are also heavily Latinx and lower income, neighborhoods such as Little Village, Brighton Park, and Gage Park (Curran and Hague 2013; see Figure 4.1). This data matches the anecdotal evidence we have of people wanting to stay as close as possible to Pilsen, but no longer being able to afford to live in the

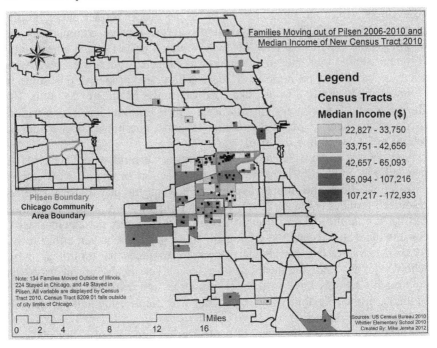

Figure 4.1 Map of where displaced Whittier families move with median income of destination neighborhoods. Map by Mike Jersha

neighborhood. This residential displacement is then used as a weapon against the neighborhood school, justifying the threat of closure and withholding of resources, resulting in an under-resourced school for those students who remain, a vulnerable population of immigrant, working class Latinxs, many of whom are learning English.

Whittier School became a site in the battle over the privatization of public education when a plan was announced in 2010 to demolish the field house of Whittier School, affectionately known as La Casita, in order to build a soccer field for a nearby private school. A group of Whittier parents, over-whelmingly mothers, occupied the building, demanding it be converted into a library. Whittier School does not have a library, just one of the over 160 CPS schools that do not have a library on-site (Curran and Hague 2013). The mothers created a volunteer-run library with donated books during the course of the 43-day occupation. CPS cut off gas service, heat and hot water to the site, claiming the building was not safe and should be demolished. The occu-pation ended with the local alderman agreeing to renovate the building with tax increment financing funds. That money never materialized. The building was demolished in 2013 with no warning. A playground for Whittier School was built in its place. While a playground in Pilsen is welcome given that the neighborhood is underserved in terms of green space, the process was undemocratic and blatantly dismissive of community concerns. As a number of parents chanted at the opening ceremony for the playground, "Dónde está la biblioteca?" (Where is the library?) (DNAInfo Staff 2014).

In this, as in other battles over school resources, Whittier, as a neighborhood school that serves a working class, Spanish-speaking population, has been pitted against institutions that seek to present a more middle class image for Pilsen. Primary among these is the charter school run by the United Neighborhood Organization (UNO). UNO is a community organization that positions itself in opposition to those who have cast "Hispanics as a victimized community in need of social justice;" rather, "UNO has worked with a population willing and able to take full advantage of American possibilities" (Sanchez 2006: 9). Similarly, the UNO educational philosophy, "seeks to redefine the culture of public education, especially in urban settings among Hispanic and African-American students. Too often, inner-city schools are satisfied attaining the low standards that many administrators and society have set for them" (UNO Charter Schools website, accessed 2010, no longer online). UNO schools require English immersion and uniforms. Students have reportedly been reprimanded for speaking Spanish in school (Zehr 2009). The goal is to distinguish the brand of UNO schools, and the students themselves, from the negative perception of public education in Chicago. Juan Rangel, former head of UNO,[1] said of their students, "Most of our kids were born here. They aren't isolated or segregated from the regular society" (quoted in Zehr 2009). The blatant othering of immigrant children and families makes clear who these new schools are, and are not, designed to serve.

The process through which schools are closed, new schools are opened, and schools are managed has become profoundly undemocratic in this neoliberal era. The decisions about school closings and new charters in Chicago are made without real participation from the affected communities. Public hearings are called with as little as three days' notice and often occur at the board's downtown office, thereby making meetings inaccessible to many community members (Lipman 2009). A former student of mine who worked for CPS told me that at community meetings about the school closures in 2013, which were increasingly heated, CPS stopped sending any high level staff and just sent a stenographer to meetings to record community complaints rather than attempting to address them in any way. Community members perceive the process as "deeply disrespectful of their wisdom about what constitutes appropriate educational decisions for their communities" (Lipman 2009: 229).

The extent of parents' alienation from the process was made evident in the summer of 2015, when a group of parents and activists went on a hunger strike to put pressure on CPS to re-open Dyett High School, the last open enrollment school in the historically African-American neighborhood of Bronzeville on the South Side. An open enrollment school accepts everyone within its catchment area. A well-developed community plan five years in the making envisioned an open enrollment school with a focus on leadership and green technology. When CPS kept on postponing meetings about the school without explanation, community members were concerned that the school would stay shut or else be handed over to a private entity. As one striker put it, "This is really about the privatization of education, it's about having sustainable community schools in every neighborhood. This is a much larger struggle" (quoted in Strauss 2015). After a 34-day hunger strike, CPS agreed to re-open the school as an open enrollment school but with an arts focus.

The resources that gentrifications draws away from schools are social as well as financial. As Lipman (2009: 227) notes, "What may appear to outsiders as 'deprived', 'run-down' and 'bad' schools have far more complex meanings for those who live there, including the web of social connectedness essential to well-being and survival." Schools serve as community power bases, because they are the place where women come together (Curran and Breitbach 2010). School closures are then a "willful destruction of social capital" (Greenbaum (2008) quoted in Lipman 2009). These neoliberal school policies increase the social capital necessary to get into "good" schools and consequently increase the stress that primarily falls on women as they attempt to secure adequate schooling for their children. The closing of neighborhood schools devastates communities and removes a means through which social capital is formed and women are empowered (Curran and Breitbach 2010). While working-class and low-income mothers will use and be active in local schools, gentrifier women use and invest resources in schools outside the neighborhood (DeSena 2006).

Public housekeeping

Jane Addams (1913) argued, "The men of the city have been carelessly indifferent to much of its civic housekeeping, as they have always been indifferent to the details of the household." By necessity, women's primary role in social reproduction has typically made women more tied to home and local concerns, with a focus on everyday responsibilities. Women's activism has tended to mirror their domestic concerns – housing, childcare, welfare, safety, the environment – a kind of public housekeeping (Garber 1995). Sharon Zukin notes, "Women have often been thrust into the role of caretaker ... Building a community, and fighting for it, naturally springs from that" (quoted in Chaban 2016). Women's activism has been central to improving urban conditions and to building social capital for their communities and themselves. Women have long done this work, but it is not always recognized or valued (Chaban 2016).

The privatization of public space and public policy has led to a "regendering" of local governance as power is transferred from the local state to various privatized and corporatized decision-making structures (Kern 2007), the Manchester Men described by Tickell and Peck (1996) as discussed in Chapter 3. This has led both to prioritizing market outcomes rather than any sort of communal care as well as reinforcing the power and vision of the men who tend to dominate such projects. This narrow group of stakeholders excludes those groups whose interests "lie outside of the regime of competitiveness" (Kern 2007: 666), often undoing advancements already achieved and undermining a community's political and social capital. In this context, formal planning processes are merely performances. Rankin and McLean (2015) describe the process of redevelopment in an inner suburb of Toronto where "creative" and "green" strategies are being sold to the white middle class while ignoring the existing community of working class immigrants. Community meetings, for example, were held in a local legion hall whose ban on head scarves precluded the participation of Muslim women wearing hijab. Racialized groups are enrolled only "tokenistically and instrumentally in the projects of elite city builders" (Rankin and McLean 2015: 230). Community meetings exist merely to say that you have done them. There is no community input into shaping what is discussed, no framework for responding to the community concerns expressed.

In the neoliberal era, social capital has been commodified as an individualistic, entrepreneurial project for developing connections that help one succeed individually at the expense of a more collective vision that brings together resources one cannot create as an individual (Lipman 2009). The result is a neoliberal consumer liberated from collective responsibility (Huse 2014). The transitory nature of many gentrifiers' residential patterns means that they do not have strong local ties and commitment to place, so civil society's capacity to provide care has lost much of its force (Huse 2014). There is a denial of state responsibility to foster equality of outcomes; family and the community are now responsible for the reproduction of good citizens (Rose 2010). Yet neoliberal housing and education policies actually serve to destroy the collective social capital in low-income communities (Lipman 2012).

One such example is the phasing out of Local School Councils (LSCs) in Chicago. In the Dyett hunger strike mentioned above, one of the central points of contention was the demand by the community that the re-opened school have a Local School Council (LSC). LSCs are elected bodies that determine how to spend school budgets and hire and fire principals. LSCs were the result of decades of struggle, particularly by African-American and Latinx communities, for equal resources, bilingual education, and against racial segregation. Made up of both parents and community members, they are the largest body of elected public officials of color in the United States (Lipman 2009). Yet, Renaissance 2010 eliminated elected LSCs in favor of appointed advisory boards or private charter school boards, further privatizing the process and undermining democracy (Lipman 2009). The work of the LSCs has been systematically undervalued, despite the direct evidence of community involvement and parent care that is so central to the neoliberal ideology of intensive mothering.

In her study on gentrification and school choice in the Greenpoint neighborhood of Brooklyn, DeSena (2006; 2009) hypothesized that one of the reasons gentrifiers select schools outside their area is that it is easier to transport children than it is to mobilize change within the New York City school system. But what if it weren't? What if gentrifiers brought their significant social capital to a neighborhood in order to organize for collective goods across class? Davidson (2012) argues that a gentrifier who engages with the existing community is a truly radical injunction. Gentrification is the articulation of difference (Davidson 2012). If we can cultivate spaces where that difference is minimized, where common visions can be fostered, we can start to unravel the unequal landscapes that gentrification requires.

Organizing around schools is one potential avenue to create these kinds of spaces. My own experience as a parent of children in CPS is that there is an incredibly skilled and dedicated group of parents (mostly mothers) organizing across class to improve not only their local school, but community schools more broadly. These parents are committed to keeping schools as diverse places and express concern about how neighborhood gentrification will affect the ability of families to stay in the area, strategizing how the school community could help with this. The parent organization Raise Your Hand advocates for equity in school resources and funding and trains and empowers parents. While this may hardly be the primary force in education policy in Chicago and other gentrifying cities right now, it is indicative of how we could think differently about care and community.

Evidence for the potential for alliances across class for community resistance to gentrification and displacement comes from my work with Trina Hamilton around environmental activism in Greenpoint, Brooklyn (Curran and Hamilton 2012; Hamilton and Curran 2013). In this case, the Newtown Creek, site of one of the largest oil spills in U.S. history, was, after decades of activism, finally declared a Superfund site in 2010, meaning that the federal government would step in to enforce clean up. Given that this designation occurred at a time that the neighborhood was experiencing gentrification, we expected to find a straightforward case of environmental gentrification, where remediation was taking place in order to serve the new professional residents and to encourage further gentrification.

That has certainly been a clear pattern in cities across the world. The provision of environmental amenities, anything from new parks to bike lanes and brownfield remediation, has been linked with real estate speculation and the displacement of the working class (see, for example, Banzhaf and McCormick 2007; Dale and Newman 2009; Dooling 2009; Quastel 2009; Pearsall 2010; Checker 2011; Dillon 2014; Sandberg 2014; Kern 2015). Indeed, Kern (2015) argues, the removal of the "dirty" bodies of the working class who are associated with the industrial past and their replacement with the "clean" bodies and practices of the new middle class allows these neighborhoods to then be coded as clean, regardless of whether they are environmentally clean or not. Often, this practice is gendered, with environmental products and services, "detoxifying, cleansing, balancing, aligning, beautifying and purifying," marketed to women, a feminization of the masculinized industrial landscape (Kern 2015: 74, and as discussed in Chapter 3).

Environmental gentrification has been possible because of a very narrow view of what constitutes the environment for some environmentalists. Calls for environmental improvements too often ignore their social justice implications. Checker (2011: 212) defines environmental gentrification as the convergence of urban redevelopment, ecologically minded initiatives and environmental activism in an era of advanced capitalism. Operating under the seemingly a-political rubric of sustainability, environmental gentrification builds on the material and discursive successes of the urban environmental justice movement and appropriates them to serve high-end redevelopment that displaces low income residents.

Calls for ecological revitalization and sustainability often ignore the history of racism and underdevelopment that have made these spaces newly attractive to development (Tretter 2013; Safransky 2014; Kern 2015). Greening and sustainability are a way of making invisible the racialized nature of this environmentalism (Anguelovski 2014). But for long-time community residents, "the urban environment is the place where people live, work, learn, and play all together. They view the urban environment in a holistic and comprehensive way" (Anguelovski 2013: 213). In this more holistic view, environmental justice cannot be achieved without paying attention to racial discrimination, housing affordability, safety, and a host of other issues that tie people to place.

A holistic view of sustainability and environmental justice should result in a deepening of democracy and in developing the tools for questioning broader political arrangements (Anguelovski 2014). While gentrification has often served to dismantle existing social networks and political power, the activists in our case study in Greenpoint were able to educate and include incoming residents in their activism and vision for the future of the neighborhood in a way that strengthened their organizing and helped to achieve results. They schooled the in-movers. When gentrifiers started new community groups, the long-term residents joined them. The movement evolved from "five angry women" to a "kick ass community" (Hamilton and Curran 2013). The long-term activists had legitimacy while the gentrifiers had useful professional skills and growing political capital. Before gentrification, "[I]t's literally just one, two, three, four, five angry women, fighting all these things like incinerators and sewage plants and having to get these things done" (quoted in Hamilton and Curran 2013: 1566). Gentrification helped to change this. As one of the 'five angry women' described it (quoted in Hamilton and Curran 2013: 1566), "People were working really hard, you know, to buy their home and to stay in their home without really realising what was happening to them...[M]any things are probably happening because of gentrification." It was only after gentrification started in surrounding areas that the contamination started to seem out of place (Hamilton and Curran 2013).

Activists were able to form alliances across class because they found common ground; industrial pollution and toxicity do not respect class boundaries. The uncertain nature of the health risks involved in the Newtown Creek oil spill (and the other industrial pollutants in the neighborhood) as well as the fact that many of the gentrifiers had been completely unaware of this industrial history before they moved there, left many people feeling duped and at risk, and therefore much more likely to be supportive of the larger social justice goals of the movement rather than solely motivated by property values (Hamilton and Curran 2013; Curran and Hamilton 2012). This creates the space to form new community identities forged in place, through everyday struggle. As one gentrifier-activist put it (quoted in Curran and Hamilton 2012: 1033), "you have real crimes being committed, environmental crimes ... your goal is to make their lives better and somehow protect them."

This does not necessarily mean that gentrification is no longer a threat to Greenpoint. New developments are planned; more hipsters have moved in. Change is inevitable, but the way change happens is up to us.

I offer the Greenpoint case study as an example of ways to contest the presumed inevitability of gentrification and to show that it is rarely, if ever, complete (Curran and Hamilton 2012). Activism and community participation can build stronger connections and pride in community (Anguelovski 2014). Public housekeeping is necessary. Though gentrification has worked to privatize and de-democratize the work done by generations of activists, it does not have to be that way. Gender is one lens we can use to see shared challenges and oppressions across race and class in a way that helps to dismantle those inequalities.

Conclusion

From parenting to community activism, care work has historically been dominated by women and devalued for that very reason. Gentrification has both reshaped and reinforced this gendering of social reproduction in ways that make care work expensive, isolating, intensive, and precarious. Gentrification has done little to reorganize urban space in such a way as to facilitate social reproduction while making the necessary services more expensive across urban space. Mothering becomes more intensive both financially and emotionally as parents compete to access the best resources for their children in ways that perpetuate existing racial and class inequalities, as evidenced in the school choice movement. Environmental activism has often been co-opted to accomplish environmental gentrification. Gentrification aims at accomplishing either the erasure or the commodification of this community work to reinforce neoliberal notions of care and social reproduction as individual responsibilities. Public housekeeping is necessary to contest this vision and can serve as an avenue for re-democratizing urban space. Without democratic community participation, gentrifying neighborhoods become less safe for their long-term residents. I turn to issues of safety in the next chapter.

Note

1 Rangel resigned in disgrace following a Securities and Exchange Commission investigation and accusations of insider contracts, nepotism and political cronyism (see Burke 2014).

5 Safety

Safety is one of the primary amenities that gentrification is selling. Gentrifying neighborhoods frequently see an increase in police presence and more aggressive policing in areas that had previously been neglected by urban policy. As William Whyte (1988) argued in his famous Street Life Project, women in the urban landscape are the canaries in the coal mine; they are taught to be good readers of urban space and safety, and thus the presence of women in urban locations indicates an area deemed safe by them. Van den Berg (2012) details how coding the city as female is part of the gentrification process, using the term genderfication to make explicit the intersectional relationship between race, class, and gender in the re-commodification of inner city Rotterdam. Cityscapes coded as female are understood as safer than what came before.

Gendered notions of fear and safety articulate well with neoliberal redevelopment agendas (Kern and Mullings 2013). Safety is something that can be purchased by moving to the right neighborhood rather than a collective good. In the context of gentrification, neighborhoods become "good" when they move towards homogeneity and "safe" when they become dangerous to their original inhabitants (Schulman 2012: 30). The gendered and racialized constructions of victims and perpetrators of crime help to fuel the narrative of gentrification and create the conditions necessary for the displacement of the existing community. This displacement is itself a form of violence.

The focus on making the city safer as a way to accomplish the gentrification of neighborhoods has focused on very particular conceptions of danger and disorder in the urban core which marginalize other areas of crimes and anxiety (Fyfe 2004; Coleman et al. 2005) and also other populations and areas of the city. This chapter explores how the selling of the safe city is based on the exploitation of women's fear of the city and of stereotypes built around young men of color in ways that act to put people at risk. Spectacle policing policies such as broken windows and zero tolerance focus on certain areas and certain crimes, rendering other crimes, like rape and domestic violence, less visible and targeting marginal populations like sex workers, rendering them less safe. In policing, as in other areas, gentrification improves circumstances only for those who can afford to participate in these newly revalued spaces. As Richie (2012) argues, women with less power are in as much danger as ever precisely because of the directions that policing has taken. I would like to suggest here, following Kern and Mullings (2013), that a recognition of the common sources behind the ways in which women are disenfranchised, policed, and constrained would provide the opportunity for new alliances across difference that counter dominant narratives of what constitutes the "safe city."

Selling safety

The perception of safety, or at least of activist policing, is essential to attracting capital to the gentrifying city (Mountz and Curran 2009) with cities across the globe adopting zero tolerance policing that targets low level crimes in order to prevent larger crimes from occurring. Despite the fact that the effectiveness of this type of policing has been widely debated, it has come to be seen as 'common-sense' in the world of policing theory (Herbert, 2001; Mountz and Curran 2009).

Policing and its effects are gendered. Herbert (2001) recognizes the existence of two different ways police "do" gender: "hard chargers" and "station queens." In this conception, the "hard charger" is seen as doing the real policing, thus reinforcing the masculinity of the street and of police work, whereas the "station queen" is assumed to inhabit the more protected space of the station. Zero tolerance plays on this masculinity, its main policy prescription being for the police to "kick ass" (Bowling 1999, p. 548; see Mountz and Curran 2009). As Alison Mountz and I detail in our work on the export of zero tolerance policing to Mexico City by former New York City mayor Rudy Giuliani, the performance of gender is key to selling the safe city. Giuliani, known as much for his multiple performances in drag as for his aggressive personality, acts the "tough-cop-as-expert" on a global stage, using his own performances of both masculinity and femininity in drag to fuel the masquerade of meaningful urban reform. Giuliani's drag appearances play femininity with masculinity, intending to make public space appear friendly and non-threatening. For example, placing women on display in the city's center as traffic cops, as has occurred in Mexico City after Plan Giuliani, perpetuates and even paints a "friendly" (read feminine) face on the illusion of control (Mountz and Curran 2009).

This performance is part of the "making up" of neoliberal policy, to mask as effective, comforting, logical, and inevitable a set of policy prescriptions that has led to more insecurity, not less (Mountz and Curran 2009). This results in the valorization of certain neighborhoods within the city, or of certain cities within the global city network, with international investors from cities with high crime rates seeking out safer locales. Satow (2015) highlights safety as one reason why the global elite want to move to New York, with one homebuyer saying, "We really want to raise our kids in a more relaxed environment, where they can be free and just walk to school without having to worry about safety." In São Paolo, for example, you would never see a high profile businessman in the market, especially without armored guards. "There you go straight from the office, to the car, to home." But the means by which city governments go about creating the image of the safe city may be dangerous to many long term residents. The zero tolerance policing celebrated as a means by which to accomplish the "cleaning up" of the city strengthens the masculinist state, one that denies civic engagement of both men and women and as such, lessens the potential for democracy (Herbert, 2001; Mountz and Curran 2009).

The use of fear as a tactic in maintaining the uneven development of urban space is an essential development tool, where notions of danger are used to legitimate revanchist policies and gentrification (Kern 2010a, 2010b). This is one of the elements employed to sell certain neighborhoods and certain developments over others, with an increasing tendency for doormen, gated buildings, and other indicators of security that separate luxury developments from the grit of the rest of the city. Indeed, these features open up spaces for development that might otherwise be deemed too risky for investment; "safety" smooths the path and speeds the spread of gentrification (Kern 2010b).

These kinds of security measures are especially marketed to women. This both feeds on the perceptions of fear, however accurate, in neighborhoods experiencing "revitalization" and reinforces the notion that women should be afraid in urban space. Kern's (2010a, 2010b) work on the gendered marketing of condominiums in Toronto, where women are at least 40 percent of purchasers (Kern 2010b), recognizes that the images used to sell this vision of the city offer both a feminized and eroticized vision of revitalization that is embedded in the nexus of both freedom and fear that fuels neoliberal urbanism. While there is a long history and literature on the struggle by women to create safe streets (Hayden 2002; Wekerle 2003), gentrification privatizes this struggle, making safety something available for purchase for those who can afford it.

A controversial ad campaign on public transportation in Chicago illustrates the intimate link between fear and women's sense of space under the guise of female empowerment. A supposedly tongue-in-cheek advertisement for pepper spray urges women to "make a grown man cry," thus attempting to profit from the fact that gender is the most significant factor related to anxiety about victimization in public transit (Loukaitou-Sideris and Fink 2009). Even as access to transit, especially rail transit, serves to make particular neighborhoods targets for gentrification, with gentrifiers trading in a car commute for access to the urban core willing to spend more money for housing close to transit (Zuk et al. 2015), there is little recognition of the way in which this access is gendered. Indeed, Loukaitou-Sideris and Fink (2009) found in their study of U.S. transit operators that though transit operators believed that women have distinct security needs, most did not think agencies should put programs in place to address those needs. So instead we are left with faux-empowering imagery that makes women feel unsafe in public space even as they are encouraged to take matters into their own hands, through their consumer purchases. Alternatively, the safety of a cab can be purchased for those who can afford it, though of course the degree to which cab service is safe is also gendered and highly place-dependent.

The benefits that many feminists have long argued accrue to women in the city (e.g. Wekerle 1984) become watered down in the drive to gentrify with newly 'safe' spaces allowing upper class white women the opportunity to become the flâneurs of the postindustrial city. Becoming part of urban areas previously deemed off limits is part of becoming authentically urban, a way through which white bourgeois identities are made through boundary crossing into spaces marked as other (Kern 2010b). But these women are themselves merely commodities, useful in how they sell the city to others and liberated only by virtue of what they can afford to consume.

Sacrificing women's bodies to sell the "safe city"

Neoliberal urban policy is happy to sacrifice these women's bodies in pursuit of the designation of 'safe' city. Brownlow (2009) shows how the entrepreneurial declarations of urban safety by city promoters today are based upon crime data that often are the product of decades of misrepresentation, namely a pervasive underrepresentation by city police of women's victimization by male sexual violence, especially rape. Rape is among those violent crimes most vulnerable to data manipulation because of rape's ambiguous definition in the criminal justice system and the geography of rape (i.e. the likelihood that it will occur in the home). Brownlow argues that the appearance of safety achieved through this data manipulation is an artefact of patriarchy. Patriarchy, then, acts as a tool of urban economic development, fabricating crime rates to achieve desired development outcomes at the expense of women's bodies and lives (Brownlow 2009).

This is made evident in the case of Shannon Schieber, a graduate student at the Wharton School who was found raped and murdered in her apartment in downtown Philadelphia. Following her murder, it was revealed that her case was only "the latest in a heretofore unidentified and unpublicised pattern of similar attacks against single, White women living in Center City Philadelphia's chic Rittenhouse Square neighbourhood – the city's principal growth pole and its epicentre of White, '20-something' urban renewal" (Brownlow 2009: 1688). Prior to Ms. Schrieber's murder, four other women had reported similar attacks to the Philadelphia Police. Similarities in the pattern and description of the crime indicated, and DNA evidence later confirmed, a serial rapist. But no warning was issued by the police, and the perpetrator raped at least 11 more women by the time he was captured in 2002 (Brownlow 2009).

Brownlow (2009) details how Ms. Schrieber's case exposed a pattern of hiding rape that went back decades, with many reports of rape categorized by the police as 'unfounded,' humiliation and indifference used as tactics to discourage women from reporting rape. When that did not work, the crime would be coded as "investigation of person" and enter bureaucratic limbo. Between 1984 and 1999 the Philadelphia Special Victims Unit mischaracterized one third of its complaints (Brownlow 2009).

This invisibilization of rape helped to construct Philadelphia as among the "best places to live" in the U.S. in the 1985 Rand McNally Places Rated Almanac. In August 1996, coinciding with the first of the Center City Rapist's known attacks, Philadelphia was ranked among the top 10 cities for corporate relocation. Young, single professionals moved to Center City in greater numbers, attracted by new developments and satisfied by the city's apparent commitment to safety. The link between the covering up of crime and economic development was made explicit by one police official, "You've got to keep the stats down. It's all about competition, because every city wants to look better" (quoted in Brownlow 2009: 1694).

As such, while the case of Philadelphia was especially egregious, Brownlow (2009) makes it clear that the problem of "cooking the books" is in no way limited to Philadelphia. The crime rate, and the illusion of safety, is socially produced in order to attract capital to the postindustrial city. But even though this "safety" was produced for upper class white women, "class and privilege are no guarantee where the politics of accumulation meet the persistence of patriarchy" (Brownlow 2009: 1697).

While even upper class white women could be victimized by the manipulation of crime data to make the city appear safer than it was, the primary victims were those women of color who were far more likely to come into contact with police, and far more likely to be ridiculed and ignored. The policy that resulted in the murder of Shannon Schieber was tested and perfected on the working class women of color of Philadelphia (Brownlow 2009).

So long term residents of gentrifying cities are left with the worst of both worlds, victims of the speculative pressure of gentrification that make it less affordable to remain in their homes, feeling like outsiders in their own neighborhoods, but not actually experiencing the benefits of reduced crime that are purported to come with gentrification. A long term activist with whom I have worked in the Pilsen neighborhood of Chicago told me, "you wouldn't believe how bad it is" (personal interview, February 2016), recounting both the continued displacement of her neighbors and local businesses as well as an increase in shootings and gang activities in the neighborhood. One such example is the accidental shooting of Aaren O'Connor, hit by a stray bullet while talking to her sister and father on the phone in her car. Police called to the scene by a 911 call reporting gunshots saw no evidence of a crime, and so her body was not found by police until a second 911 call from her roommate (Lulay 2016). Someone responded by spray painting "Aaren O'Connor's killers" on the side of a house in Pilsen (Lulay and Nitkin 2016). Media coverage of the crime found special tragedy in the fact that O'Connor had just recently moved from California to be with her boyfriend, who lived in

Pilsen, but not that the case is part of a larger increase in gun violence experienced in Chicago in 2016. This, despite the fact that an ongoing argument for gentrification in Pilsen (and beyond) has been that it will result in the reduction of crime. Those who have resisted the gentrification of the neighborhood were presented as complacent residents who have no problems with the crime and drugs that plagued Pilsen through the 1970s and 80s. In this way, pro-gentrification forces "Use the past as a weapon against the present" (Wilson, Wouters, Grammenos, 2004: 1182). When this same community activist, a life-long Pilsen resident, expressed her opposition to a condominium development to a campaigner for the reelection of the area's Alderman, Danny Solis, the campaigner asked her, "You want to live in the ghetto?" This is the starkness of the choice presented to urban residents: gentrification or ghetto. Yet the reality is that people are under siege on a number of fronts, insecure both in their housing tenure and their personal safety, in ways that are distinctly gendered, from the perceived vulnerability of female bodies to the scripted danger of young male bodies of color.

Spectacle policing

On the flip side of "cooking the books" to make the city appear safer is the creation of spectacle policing. Brownlow (2009) summarizes the formula as such: inflate street crime to a level of crisis; allow the local media to stoke public fear and generate public support for police actions; and, respond to street crime in often-sensationalized, 'made for media' fashion, thereby demonstrating police competence. This form of policing is just as gendered as cooking the books, in this case targeting and stigmatizing young men of color.

David Wilson and Dennis Grammenos (2005) detail the process in Humboldt Park, a gentrifying Puerto Rican neighborhood in Chicago. They argue that real estate capital and the media have targeted and scripted Puerto Rican bodies, particularly young male bodies, for a new "gentrification-sanitizing theme: a disgust for 'ghetto' morals and social order" (Wilson and Grammenos 2005: 296). The area "was being modernized and socially re-made to tackle the area's youth problems and cultural dilemmas" (planner from Chicago's Department of Planning and Development, quoted in Wilson and Grammenos 2005: 302). The story told of the neighborhood by one developer was as such, "To understand Humboldt Park and what it needs, you must know about these kids. These are different kids … they walk the streets oblivious to its problems and dangers and meet up with others for fun and amusement … they lose themselves in the street world … Gangs, too many gangs" while another commentator argued, "The young men congregat[e] on the corners… You have to be fearful of them" (quoted in Wilson and Grammenos 2005:302). Coded as glaring, intimidating, ignorant, foul-mouthed, baggy pants and baseball cap-wearing, and non-English speaking, Wilson and Grammenos (2005) argue that the message from this discourse is clear:

these kids and the parenting styles and social relations they represent, could be confronted, and corrected, by gentrification.

Similar coding work is on display in the remade communities of the former public housing developments in Chicago discussed in Chapter 2, where the nature of community and the rights and responsibilities of the members of those communities are highly contested. Racial assumptions about the characteristics of public housing residents, and renters in general, are reproducing the class and racial divisions associated with the stigmatization of public housing that redevelopment was supposed to improve. Instead, new forms of exclusion and marginalization are being enacted in these communities, with increasing surveillance, control and rule enforcement as the new community tries to define what the behavioral norms should be (Chaskin and Joseph 2015). Following the "broken windows" narrative conflating crime and incivility, behaviors like playing loud music, hanging out, arguing in public, littering, and even storing personal items on balconies or hanging laundry in view become violations which can lead to eviction. Three lease violations result in eviction. As one development professional stated, "We have to rack up the lease violations … if you breathe hard, that's a lease violation" (quoted in Chaskin and Joseph 2015: 189).

This is part of a larger trend in which gentrification results in the progressive criminalization of "quality of life" issues and the tendency to censure what are legal behaviors (Pattillo 2007). In these neighborhoods experiencing redevelopment, the very presence of African American men is seen as an indicator of drug or other criminal activity (Chaskin and Joseph 2015). This criminalization starts early. In one of the redeveloped projects in Chicago studied by Chaskin and Joseph (2015), children as young as 10 and 11 were arrested by a security guard for being too loud and too old to play on the playground. The stigma is such that a resident at one development suggested armbands to identify resident youths (Chaskin and Joseph 2015).

The stigmatization of urban areas associated with young men of color and criminality is essential to securing whiteness (Shaw 2007) in the urban core. In her work on the construction of Sydney, Australia as a city of whiteness, Shaw details how the urban problems experienced by Aborigines, especially in the Sydney area known as "the Block," have been used as an indictment of all that is "wrong" with Aboriginality (Shaw 2007: 48). In the Australian imagination, Aborigines are supposed to be tribal and nomadic, thereby rendering them "foreign" in the city (Shaw 2007: 139). This performance of whiteness requires constant surveillance, with monitoring of the Aboriginal community by police, white residents, and the mass media continually feeding the narrative of Aboriginal decline on the Block, "the most surveilled area in the country" (quoted in Shaw 2007: 58). Surveillance takes the form of CCTV, fortressing, back to base alarm systems, private security firms, and extensive communication between white residents and the police that contribute to a sense of being under siege. It does not take much under these circumstances for any incident to be dubbed a "riot." (Shaw 2007: 59). This surveillance is a necessary precursor to securing what Shaw (2007: 55) calls the "whitewish," the desire for exclusionary gentrification. Attracting investment from the potential market of gentrifiers requires the reduction of Aboriginal housing on the Block. In 1968 some 35,000 Aboriginal people lived in Redfern, but by 2011, the time of the last census, it was at just 300 (Teece-Johnson and Burton-Bradley 2016). The area is being redeveloped, demolishing social housing for the Aboriginal community, to be replaced with a mixture of public and private housing. The narrative that accomplished this displacement ignores the fact that one of the central reasons for the historic concentration of the Aboriginal population in the Redfern neighborhood in which the Block is located was that it served as a focal point in the 1930s for Aborigines attempting to reunite with kinfolk from whom they had been separated by numerous forced separations and displacements (Shaw 2007).

Surveillance as a strategy for displacement is on display in the DePijp area of Amsterdam where, de Koning (2015) argues, gentrification is a state policy aimed at attracting whites to neighborhoods seen as black within the context of the privatization of social housing. City policy focuses on young white urbanites rather than native Amsterdammers who are low-income and from migrant backgrounds. Young "Moroccan" men are coded as "troublesome and fraudulent" (de Koning 2015: 1214), involved in anti-gay and anti-Jewish activities, which further serves to indict their parents, and indeed, the larger culture of the neighborhood. A response to criminal activity among such groups is to sell off social housing units, a policy that has over time resulted in a 20 percent reduction in the supply of public housing. This focus on youth further serves to indict their parents, overwhelmingly mothers, and to delegitimize their right to city (van den Berg 2013). Security enforcement further acts as a tool to displace these populations seen as undesirable, with CCTV and intense policing used to make urban areas hostile to young men, especially those of Moroccan descent. Young men deemed most undesirable on police lists are subject to special surveillance, monitoring of their contacts, and frequent write ups, so much so that many men wanted to leave the neighborhood in order to escape the surveillance (de Koning 2015).

Young female bodies of color also serve as key sites in the struggle over urban space (Cahill 2006). Cahill (2006, 2007), in her work organizing young working class women of color on the Lower East Side of Manhattan, details how stereotypes about young women of color as lazy and on welfare, likely to become teen moms, promiscuous, uneducated, in abusive relationships, and burdens to society are used in the discourse of gentrification to justify disinvestment from these young women and facilitate their displacement. Disinvestment from public housing, public education, libraries and related projects increases stress and creates unhealthy environments. This disinvestment is experienced as violence; it is about regulating young women's agency (Cahill 2007). Internalizing of these negative stereotypes makes the young women feel like they are out of control and living in hostile territory, that they must self-regulate in order not to fulfill these stereotypes. These gendered and racialized representations "reference historical caricatures of the 'tangle of pathology' and the underclass that blame young women of color for the poverty of their community" (Cahill 2007: 214).

This pathologizing of young women is necessary to accomplish their effacement from the neighborhood. Aside from the negative stereotypes, there is simply no place for young working class women of color as they actually exist in these neighborhoods undergoing gentrification. As one of Cahill's (2007: 215) researchers comments, young women are not "getting attention – and what I mean by that is that they're not thought of [...] they're just not considered. There's no space made. They're not considered for anything at all." Rather, stereotypes stand in for real people and act as critical tools in securing the consent of the public for gentrification and displacement, so that displacement is seen as inevitable and even a sign of progress (Cahill 2006).

Seeing black and brown bodies in this way is an exercise in power (Wilson and Grammenos 2005). This coding scripts not just individual bodies, but entire neighborhoods. And, Wilson and Grammenos (2005: 297) argue, "with identity and community negatively defined, prospects to present gentrification persuasively to the public and city as colonizing and unfair to inhabitants are damaged."

Wilson and Grammenos (2005: 308) call anti-gentrification forces to task for ignoring this assault on the youth body, arguing that their failure to defend youth bodies in Humboldt Park allowed real estate capital to control their use and thereby provide a vision of the Puerto Rican community that frames it as a civic problem in need of removal. This framing makes gentrification appear "progressive and city-serving." These place-making practices serve to normalize and naturalize the distribution of people and the changing character of places becomes common sense (de Koning 2015).

The Black Lives Matter movement has emerged as one way to contest the stigmatization of these youth of color. Black Lives Matter (BLM) started in the U.S. as a response to police brutality, but has evolved into a more global effort to combat racism (Tharoor 2016). The movement, founded and largely led by women, recognizes the ways in which Black people are intentionally left powerless at the hands of the state (http://blacklivesmatter.com/about/), from police brutality to the destruction of public housing and the displacement of communities of color through gentrification, and that black women bear the brunt of these multiple oppressions. In this way, BLM has made the connection between gender and gentrification in a way that most of the literature on gentrification has not. The safety of young bodies of color in public space is the gateway through which to build a larger movement for a right to the city. The very fact that we need to say that black lives matter shows there is still a long way to go.

Another group specifically targeted for displacement in an effort to construct the safe city is sex workers. In cities around the world, the removal of sex workers from public space is a key strategy in zero tolerance policing. These campaigns are one of the ways that gentrification serves to recenter masculinity in the cityscape at the same time that it encourages capital accumulation (Hubbard 2004). Hubbard (2004) argues that neoliberal policy reinscribes patriarchal relations onto the city, serving both capital and the phallus. The goal here is not necessarily to end or fix the problem of "vice." It is simply to make it less visible and to open up areas of the city for capital investment. Indeed, Papayanis (2000) argues that the removal of sex shops, brothels and massage parlors is a necessary precondition for the resurgence of central city real-estate markets. This has occurred in cities across the globe, from London, Paris and New York (Hubbard 2004; Papayanis 2000) to Seoul, where the city's urban renewal plan explicitly recognized sex workers as "irreconcilable with the city's middle class ambitions" (Cheng 2013). When sex workers and other evictees in Seoul's Yongsan red-light district, targeted for demolition, protested to demand fair compensation to relocate, the police launched a violent crackdown that resulted in a fatal fire in which four evictees and one policeman were killed. "[T]he progress of the city depended on purging the spaces in which these individuals lived and worked" (Cheng 2013).

This displacement serves both to valorize the "appropriate" heterosexual family while also creating a restricted sexual economy from which capital can profit (Hubbard 2004). The burden of the "clean up" is gendered, with prostitute women suffering most acutely from revanchist policing, forcing them to work in clandestine, and therefore precarious, conditions (Hubbard 2004). As a consequence, Hubbard (2004: 677) notes, "(male) pimps and managers have actually profited from spatial displacement, with street workers increasingly seeking employment in escort agencies or off-centre massage parlours rather than risking working independently."

Melissa Wright (2014) links this urban cleansing to the creation of a landscape of assassination, in which the city, in this case Ciudad Juarez, and much of the population who built it must be destroyed in order to repopulate the city for purposes of gentrification. Female sex workers are among the first groups targeted, coded as dirty whores rather than as independent business women making rational economic choices for better pay and more flexibility, especially in regards to childcare, than would be available in the export processing factories known as maquiladoras. The city center had once functioned as a place of safety for these women, a place where they could look out for each other and did not have to pay a pimp any fee. Plans for the "clean up" of the historic center ignored these women's role as legitimate and contributing members of society and instead coded them as invaders who had destroyed a previously pristine social fabric that had never, in fact, existed (Wright 2014). The removal of this population then becomes an indicator of progress. As Wright (2004: 371) argues, "By representing what value is not, she establishes what value is. In her opposition, therefore, we find value's positive condition. And following this logic, we find progress in the places where she once worked, in spaces she once occupied, in the city she once inhabited."

The logical conclusion to these narratives is the feminicidio: the killing of women with impunity in Ciudad Juarez in the 1990s. The official portrayal of the victims as prostitutes creates a form of social cleansing, in which these women are seen as not having legitimate claims to belonging in the first place. This script is then replayed in the social construction of the victims of the drug war, with working class youth caught up in the violence coded as dangerous invaders rather than as long-term residents of the city, the children of the women targeted in earlier social cleansing campaigns (Wright 2014). These examples demonstrate the violence it takes to bring private property markets into being and to maintain them (Safransky 2014).

In danger at home

"Cleaning up" the city and the concomitant rise in housing prices makes women vulnerable on (at least) two fronts: the danger of displacement and the danger of domestic violence and sexual harassment, which women may be forced to endure in order to stay housed. Domestic violence has long been a primary cause of homelessness for women and their children (Arnold and Slusser 2015). In an environment of scarcity of both social services and affordable housing, women and others remain trapped in abusive relationships so as not to lose housing, especially following the widespread closure of shelters for victims of domestic violence following the 2008 financial crisis (Barry 2015). This is all the more ironic given that the term zero tolerance was first coined by an Edinburgh women's domestic violence campaign in relation to violence against women (Fyfe 2004).

Nuisance laws have been adopted by many cities in a turn towards community policing since the 1990s designed to reduce fear, disorder, and incivility. These laws fine landlords for repeated, "nuisance" calls to 911. Research shows that these laws have had an unfortunate impact on women who are victims of domestic violence, for whom repeated calls to 911 resulted in threats of, or actual, eviction. A history of eviction prevents tenants from obtaining affordable housing in a decent neighborhood and it disqualifies them from many housing programs (Desmond 2016). Nuisance laws discourage victims from calling the police for protection, exacerbate the barriers that victims already face in securing housing, and unfairly blame the victim for criminal activity that she cannot control (Arnold and Slusser 2015). A 2013 study in Milwaukee found that the majority of tenants threatened with eviction were battered women rather than the batterers and that as a result of downgrading battered women's 911 calls from a potential crime to a nuisance, many landlords concluded that domestic violence was "petty, undeserving of police protection" and that landlords "assigned to battered women the responsibility of curbing the abuse." The laws, therefore, force women to choose "between calling the police on their abusers (only to risk eviction) or staying in their apartments (only to risk more abuse)" (cited in Arnold and Slusser 2015).

Women are also more susceptible to sexual harassment in order to maintain their housing. "Special arrangements" can be made (Muñiz 1998; Curran and Breitbach 2010). Female tenants experience unwanted sexual comments and advances, offers of reduced rent for sexual favors, landlords who expose themselves, and threats when sexual advances are rejected. Tenants often feel their only choice is to endure the harassment or face living with their children on the street (Mock 2015b). The shrinking of the affordable housing market that gentrification creates results in the increased vulnerability of working class women to sexual harassment form landlords.

The end of neighborhood as refuge

With all the focus on crime statistics and policing strategies, it is easy to forget that the feeling of safety and belonging is much of what makes a neighborhood feel like home. The narrow definition of harm employed in zero tolerance and similar policing strategies forecloses scrutiny of the city-building process itself, ignoring the harms generated through urban neoliberalism (Coleman et al. 2005). The displacement engendered by gentrification endangers the sense of safety and refuge that urban residents have constructed for themselves. Gentrification leads to a "fear of erasure" from the neighborhood and the city (Anguelovski 2013: 223) that leaves people feeling a need for protection and nurturing. Anguelovski (2013; 2014) builds upon Fullilove's (2004) concept of root shock, where urban renewal triggers feeling of amputation, and the loss of place has devastating consequences for memory and identity.

Shaw and Hagemans (2015) detail the ways in which every day displacements affect low-income residents in a neighborhood in Melbourne and find that the loss of familiar services like shops, meeting places, neighbors, and the nature of local social governance is profoundly disruptive, leading to a loss of sense of place without any means with which to find a new one. They find this dislocation to be just as distressing as physical relocation, profoundly affecting residents' sense of safety and belonging, even in a local context in which their housing affordability was secure. This particularly affects those most dependent on local services such as senior citizens, women with children and other caretakers, and recent immigrants.

Immigrant women are vulnerable on a number of fronts. Huse (2014) finds in her study of gentrification in Oslo, Norway that the tight knit social networks of the city center are especially important to immigrant women who might otherwise be isolated in their homes. The city center has a concentration and diversity of services necessary for women, such as, in her study, a women's shelter for teenage girls seeking refuge from arranged marriages. Pakistani girls reported feeling safer in dense urban neighborhoods where there is a greater chance someone will hear and report family conflict. Their social networks allow them the opportunity to mediate for one another. The gentrification of their neighborhood disrupts these networks and makes these women feel less safe. The increased policing that often accompanies gentrification also creates anxiety for immigrant women. Sweet (2016) finds in her work with Mexican women in the U.S. that they feel vulnerable around the police, both because of their politically constructed immigration status and because of a fear of sexual assault.

A feeling of belonging is central to feeling safe. This has been taken from working class residents in gentrifying neighborhoods in ways that are distinctly gendered, with male bodies coded as dangerous and female bodies coded as at risk. As Older (2014) argues, "Gentrification is violence. Couched in white supremacy, it is a systemic, intentional process of uprooting communities … [T]he central act of violence is one of erasure. Accordingly, when the discourse of gentrification isn't pathologizing communities of color, it's erasing them." Desmond (2016: 98), in his study of evictions in Milwaukee, sums it up this way, "Poor black men were locked up. Poor black women were locked out."

The Black Lives Matter movement has made the connection between the aggressive policing and criminalization of youth of color and the systematic removal of black people from desirable inner city locations. The movement has expanded beyond its original focus on police brutality to look at issues such as housing affordability, recognizing that the stigmatization of poor inner city residents leads to their criminalization, the targeted redevelopment of the places where they live, such as public housing, and thus, their eventual displacement. As one Black Lives Matter protestor in Cambridge, MA put it, '[Gentrification is] the erasure of black people, the erasure of people of color, the erasure of immigrants, of all poor people' (quoted in Gurley 2016). This is accomplished in the name of making the city safer.

Conclusion

While proponents of gentrification often cite a safer neighborhood as one of the benefits of gentrification, this understanding of fear and safety is both limited and distinctly gendered. I have tried to show in this chapter how gentrification is itself a form of slow violence (Kern 2016), stigmatizing both men and women of color, turning physical safety into a luxury for the few who can afford it, while exacerbating the vulnerability in both the public and private spheres of those who cannot afford it. The assumption of safety in gentrified neighborhoods is a performance, a means by which to clear urban neighborhoods of existing residents to allow for new investment. Building alliances across all those communities disadvantaged by this limited view of safety, as Black Lives Matter attempts to do, presents an alternative vision of what it means to be safe in the city, one that includes the right to the city, the right to stay put.

6 Queer spaces

Though there is a strong link in the popular imagination between gay spaces and gentrification, the connection is a spurious one. The claim to space that the gayborhood represents is the result of historic oppressions (Lauria and Knopp 1985; though oppression is not enough to form a gayborhood, see Visser 2013). The fact that some of these spaces have been colonized by capital is a function of this stage of late capitalism, not evidence of wide-spread cultural acceptance. We also need to recognize that the work on the gayborhood goes only so far in explaining the lived reality of a diverse com-munity in the Global North and completely ignores the experiences in the Global South (Visser 2013)

Gentrification is not queer. Although there is a well-established literature on gay neighborhoods and gentrification (e.g. Lauria and Knopp 1985; Binnie 1995; Nast 2002; Collins 2004; Binnie and Skeggs 2004; Brown 2014), this work is largely focused on the emergence of white, middle class, homo-normative, gay male neighborhoods that have been so marketed and main-streamed that these "gayborhoods" now face erasure (Brown 2014; Ghaziani 2014). As with urban spaces generally, queer spaces have largely been defined by homogeneity and segregation (Martinez 2015), narrowly defined by class and race and also by gender. Chicago's Boystown, for example, emerged in part because local landlords considered it more attractive to rent to gay men than to the Latinx families that had previously dominated the area (Betancur and Smith 2016).

Socioeconomic disparities are reinforced in LGBTQ communities, particu-larly among women, people of color, the young and the old, and transgender people, indicating that while homosexuality if more visible and accepted, overlapping systems of oppression exacerbate the marginalization of LGBTQ people (Goh 2015). The emergence of gay neighborhoods is linked to histor-ical forms of oppression to which spatial concentration was a response, a strategy to establish gay culture and power (Castells 1983; Lauria and Knopp 1985). While successful in that respect, they have also explicitly marked spaces as male, often ignoring lesbian contributions and excluding non-white, immigrant and trans people from LGBTQ history (Doan 2007), resulting in what Schulman (2012) has called the gentrification of the mind.

As Hanhardt (2013: 188) argues, in today's cities, marginalized identities can function as markers of cultural value but cannot be considered as vectors of exploitation. We celebrate difference by ignoring the structural constraints that create and solidify it.

In this chapter, I detail the ways in which what Nast (2002) calls queer patriarchy has reinforced the gendered, classed, and raced divisions of urban space. I explore how gentrification has been both the product of the creation of gay spaces and a tool in their dissolution, indicating that however mainstream gay culture and policies like gay marriage have become in certain geographic contexts, the ultimate goal of neoliberal policy is co-optation rather than liberation, with gay spaces marketed as cosmopolitan spectacle (Binnie and Skeggs 2004; see also Papadopoulos 2017). But this is hardly inevitable. I offer alternative examples of, and ways to think about, queer space that get us closer to the utopian futurity of queerness (Muñoz 2009).

Queer patriarchy/ queer absences

While the gayborhood is often understood as a post-Stonewall phenomenon, the association of particular neighborhoods with gay populations dates (at least) to the Chicago School of urban sociology. The Chicago School's association of homosexuality with particular urban places was so complete that by 1938, Burgess expected his students to answer in the affirmative this true-false exam question, "In large cities, homosexual individuals tend to congregate rather than remain separate from each other" (quoted in Heap 2003: 467).

Indeed, the Chicago School may be at least partly responsible for many of the silences we continue to experience in our understanding of queer urban space. In associating a particular neighborhood with an easily identifiable population, in this case gays with the Near North Side of Chicago, researchers overlooked lesbians and gay men who lived and worked in other areas of the city and, as such, overestimated the uniformity and cohesiveness of queer experience and identity. Thus, in focusing on the public world of gay Chicago, they ignored larger social forces, such as women's restricted access to public space and disposable income, racial segregation, and middle class economic and social imperatives that encouraged gays to live more closeted lives (Heap 2003).

We still engage in this overestimation of the cohesiveness and uniformity of gay neighborhoods, most notably with the rise of the creative class and Florida's (2002) assertion that tolerance for the gay community was a key marker of cosmopolitanism and therefore essential to the creative city. In Dublin, Ireland, one line of argument in support of the gay marriage referendum in 2015 was that its passage would signal Dublin as a tolerant place, welcoming of the creative class, and therefore open up new development and tourism opportunities (interview with Philip Lawton, July 2016). As a result, the twin brothers from Northern Ireland who opened a controversial cereal café in the

Shoreditch neighborhood of London now have their sights set on Dublin. The cereal café was the target of anti-gentrification protests for selling bowls of cereal for between £2.50 and £4.40 in a high poverty neighborhood. The brothers, who are gay, explained their choice of Dublin rather than their native Belfast by saying, "if Belfast isn't ready for gay marriage, then it isn't ready for a cereal café" (quoted in Boyd 2016).

Gay-ness has been linked with gentrification in a powerful way, both in the public imagination and in policy circles. But the way in which gay villages have been marketed and promoted serves to enhance the cultural capital of the already privileged, namely middle and upper class white men, at the expense of those who do not fit that image (Binnie and Skeggs 2004). As Nast (2002: 880) puts it, "certain gay men have been *colonized* by the market … Moreover, many elite (mostly white) gay men have access to the means by which to consolidate and shore up previous rounds of patriarchal white privilege accumulations." So, gay social and political movements have come out of the closet, but "usually within racist, sexist and pro-capitalist discourses … [T]hese scenes have been developed primarily by and for white middle-class markets" (Knopp 1995: 158) in cities such as Amsterdam, London, San Francisco and Sydney.

Lesbians, on the other hand, are rarely addressed as consumers, rarely targeted as a specific group and rarely identified as cosmopolitan (Binnie and Skeggs 2004). Women have not had the same ability to claim place and make territorially distinctive neighborhoods (Gieseking 2013) and when they have, these tend to be short lived (Podmore 2006). Lower incomes among lesbians make it harder to own property, while the fact that women are more likely to have custody of children affects both what they can afford and what amenities they seek (Doan 2007). The gentrification of industrial and older commercial areas in which lesbian bars had traditionally located means that lesbian bars have been pushed out as condo conversions and other speculative real estate practices take hold (Wolfe 1992). Gentrification, according to Wolfe (1992), is wholly detrimental to lesbian life (Rothenberg 1995). Indeed, as Gieseking (2013) argues, for many lesbians in her study in Park Slope, Brooklyn, the very nature of territory-making was politically questionable, a project of patriarchal colonization that produced exclusive spaces from which they themselves had been rejected.

Podmore (2006) details the rise and subsequent dispersal of the lesbian neighborhood of the Plateau in Montreal to show how lesbian spaces have been subject to change from both real estate markets and changes in queer politics and identity. While sex segregation in gay spaces in the 1980s led to a proliferation of lesbian-oriented spaces where women could be safe from the harassment of men, the 1990s saw the rise of two developments that diluted this claim to space. The first was gentrification of the Plateau. What attracted lesbians to the neighborhood were the same amenities that would later draw in the gentrifying middle class. In the Village, the gay male enclave, by contrast, gentrification was driven by the gay population. Indeed, lesbian enclaves may be gentrified by gay males, as seems to be the case in Andersonville, also known as Girlstown, the lesbian enclave in Chicago. There, gay male couples are replacing lesbians as rents rise on both residential and commercial spaces, leading to the closure of long-time lesbian-owned businesses (McGhee 2016).

The 1990s also saw the rise of a stronger identification with the common cause of a queer community among lesbians, both in Montreal and more broadly. While this potentially meant an increase in territory as the queer spaces in which lesbians were included grew dramatically, the growth of mixed bars and clubs resulted in the deterritorialization of lesbian bar culture and the consequent loss of lesbian territory (Podmore 2006). This affects lesbians as well as transgender people, for whom lesbian bars often served as safe spaces (Namaste 1996). The result is that lesbian nightlife revolves around temporary venues such as women's nights in gay bars or in community sites, something that holds true for many U.S. cities as well. These are advertised through word-of-mouth and e-mail lists and thus may not appear in official "queer" guides. Thus, as elsewhere (see Binnie and Skeggs 2004 on the scene in Manchester, U.K.), attempts to claim and consolidate queer space brought gendered results (Podmore 2006).

While gay neighborhoods are certainly more welcoming spaces across the gender and sexuality spectrum, they are masculinely gendered (Doan 2007). Valentine (2007), in his ethnography of the category transgender, argues that the politics of gay gender normativity has been shaped by the desire to delink male homosexuality from femininity. This separation of gender and sexuality has therefore required a new category, transgender. As such, "gay" has been constructed "on a model of (male) gender normativity, which impoverishes understandings of systematic gendered structures central to lesbian subjectivities; relies on a rejection of racialized public gender variance/sexuality; and implicitly elides other possible organizations of gendered/sexual experience" (Valentine 2007: 238–9). Gentrification of the gayborhood has been part of achieving both the masculinization of urban space and the privatization of sexuality, allowing inclusion of homonormative gays and lesbians into the neoliberal economy, "in which individual rights are as closely associated with market participation as they are with any theory of civil liberties" (Valentine 2007: 241).

One such example of this conflict playing out in space is the gayborhood of Chicago, known both colloquially and in city planning documents as Boystown, "suggesting that the fundamentally masculine character of urban politics was reproduced in its gay variate rather than displaced" (Stewart-Winter 2016: 9). The area's main commercial strip, Halsted Street, is marked, as Nast (2002: 884) puts it, by a half-mile stretch of 23-foot-tall "phallic bronze pylons, clad in rainbow cock rings ... The phallic rimming of Boys Town speaks of a masculinist virility, strength, and control" (see Figure 6.1). It also spatialized gay identity "in a way that linked gayness and whiteness powerfully and permanently," securing the loyalty of gay politicos in an era of gentrification (Stewart-Winter 2016: 221). The focus on this North Side

Figure 6.1 Center on Halsted, Chicago, with rainbow pylon in front

neighborhood ignores the black South Side, presenting the experience of Boystown as universal, rather than one nuanced by race and other factors (Betancur and Smith 2016). The pylons serve as a claim to space and feature prominently in real estate marketing. These kinds of overtly sexualized and gendered spaces can be alienating to women and transgender people and are spaces in which rigidities of class and lesbian and gay identity are reproduced (Binnie and Skeggs 2004). One Boystown bar, Wang's, was investigated for banning women after 11 p.m, part of a larger trend in which women are made to feel unwelcome in gay male bars (Erbentraut 2012).

Yvette Taylor's (2013) work on the dislocations and disappointments felt by working class lesbians in the UK highlights the gendered and classed exclusions created in gentrified gay villages. The gendering of most queer spaces as male left lesbians feeling out of place and "old fashioned" for wanting women-only spaces free of the "pretty little boys" who dominate the scene. Gay men were char-acterized as more individualist and hedonistic, with gay women starting to emulate that in a way that "rubbed the politics out of it" (quoted in Taylor 2013: 170). Internalized sexism within the scene continues, judging and objectifying women based on their appearance. Many also lament the expense of coming out, of participation in the scene being predicated on their ability to afford to con-sume it, with a trade-off between the politicized occupation of bad space or the apolitical consumption of trendy places (Taylor 2013).

The gentrification that frequently results from this commodification of queer culture makes these spaces that are so necessary inaccessible to those who need them most. Queer spaces serve as the entry point for transgender people. In Doan's (2007) survey on transgendered perceptions of urban space, she found that trans people feel significantly safer in cities with a visibly queer area than in those without such areas. Yet, when asked why they do not live in these areas of the city, the most frequent response was that they were too expensive (Doan 2007).

The modern gay village is thus rife with exclusions. Even the act of identi-fying the gay village has served to cement the relationship between a narrow view of gay identity and privatized city space (Hanhardt 2013). Tensions erupt when the neoliberal gayborhood is forced to contend with the "queer unwanted" (Taylor 2013). This is evidenced by the changes in what many consider to be the birthplace of the gay rights movement, Greenwich Village in New York, the location of the Stonewall Inn. While high property values have rendered the neighborhood mostly white and wealthy, at weekends public spaces are dominated by black and Latino queer youth. But, in a place where the essential authentic branding ingredient is sexuality, race has no place (Binnie and Skeggs 2004). These youth are under threat of eviction with the frequent police harassment of African American gay bars and arrests of trans women on charges related to prostitution. This is ironic given that the uprising at Stonewall that launched the gay rights movement was dominated by racial minorities with a strong transgender contingent (Andersson 2015). Research shows that when queer spaces were raided by the police, it was the "gender outlaws" who were attacked most vigorously (Namaste 1996: 231). Andersson (2015; see also Hanhardt 2013) details how the increasing concern with property values and so-called quality of life issues in Greenwich Village has led to the targeting of the people who need the space most, queer youth of color who have few safe spaces in the neighborhoods in which they live, employing the same tactics that led to the Stonewall uprising in the first place. Minority gay bars remain key sites for the city's regulation of queer space, and disorder that occurs in the neighborhood is routinely linked to "transient users," youth of color from uptown, the outer boroughs, and New Jersey. Youth are attracted to the Greenwich Village in part because there are no equivalent spaces in their own neighborhoods, as evidenced by the crumbling Harlem waterfront which continues to suffer from disinvestment (Andersson 2015).

These youth are frequently animalized and labeled as wild by both media and local homeowners groups, and interestingly, often gendered as female, with images of "bloodthirsty young lesbians" and a "wolf pack of lesbians" who serve as threats to white masculinity and the white home (Andersson 2015). Trans women are particularly targeted, with police assuming, based on style of dress and mere presence on the streets, that trans women are engaged in prostitution. This targeting has become so pervasive in New York City that the Legal Aid Society has filed a federal civil rights lawsuit (Bellafante 2016). While the organizations at the forefront of the movement to displace queer youth of color are not gay organizations, they do have gay members. The goal fits neatly with the neoliberal "vanillaization" of urban space (Andersson 2015: 276), a project of both racial and sexual purification. Thus, the very people who are producing queer space are those targeted for displacement (Andersson 2015). The concentration of wrongful arrests of trans women in New York neighborhoods undergoing rapid gentrification, such as Bushwick in Brooklyn, indicates that the purpose of these arrests is to make these women leave, thus clearing the way for even more intensive gentrification (Bellafante 2016).

A similar dynamic is playing out in Chicago's Boystown neighborhood where a community center that serves queer youth, many of them youth of color from the city's South and West sides, has generated controversy over policing and property values. The Center on Halsted is located on the stretch of Halsted highlighted with the rainbow pylons mentioned above, at the center of the gay commercial strip in a highly gentrified, overwhelmingly white upper middle class neighborhood. The Center on Halsted shares its location with a Whole Foods (see Figure 6.1). Local homeowners, both gay and straight, complain that the center attracts "problematic" youth who commit crimes and harass residents (Sosin, 2012). A jump in crimes in the summer of 2011, including a stabbing caught on video, led to calls for greater policing of the area. Politically, the increased visibility of police is dangerous for the creation of a truly queer space, preempting networking with the homeless, sex workers, and transgender people, communities who are continually harassed by police (Namaste 1996). Gay gentrifiers, hailed as a remedy for urban problems, are coded as at risk by a criminal element understood to be the racialized poor (Hanhardt 2013). The controversy resulted in the formation of a Facebook group, "Take Back Boystown," and other online content that blames queer youth of color for the crime in the neighborhood. This opposition has been criticized for its classism, racism, misogyny, femmephobia and homophobia (Lang 2012). Youth who use the Center's services complain of strict rule enforcement, the increased policing of what had been communal areas, and the frequent banning of youth resulting in decreased access to services at the Center. One youth commented, "They get funding for youth, they take the money, and then they ban the youth" (quoted in Sosin 2012). A study conducted by the *Windy City Times*, a gay publication, found that of the over 100 youth interviewed, the majority of them had been banned from the Center on Halsted at some point (Sosin 2012).

The fact of the concentration of resources for the queer community like the Center on Halsted in Boystown and similar enclaves in other cities, while often driven by the need for safety and a desire to reach their target audience, can have the perverse effect of making these enclaves the only safe spaces for the LGBTQ community. The density of LGBTQ-oriented services in the predominantly white male gayborhood draws these limited resources away from smaller organizations around the city, especially in communities of color. As gentrification in these gayborhoods leads to an increasingly racist and transphobic backlash, even these spaces may no longer feel safe (see Melt 2015).

So, the fact of gay neighborhoods does not necessarily lead to more inclusive, queer spaces (Doan 2007). As Lauria and Knopp (1985) recognized, the participation of gays in the revalorization of physical neighborhoods has been strategic in terms of developing some political and economic power, but may not serve the human needs of individual gays. Schulman (2012) relates this gentrification of physical space to a spiritual gentrification, the gentrification of the mind. Her thesis (Schulman 2012:15) is that a "certain urban ecology of queer subcultural existence has been wiped out, through both AIDS and gentrification; this ecocide has resulted in less diversity." She recounts how the AIDS crisis literally made the space for gentrification, with the deaths of her neighbors in Greenwich Village rapidly turning apartments over to market rate. A glaring example of this shift, and of the erasure of gay history and experience, is the fact that St. Vincent's Hospital in Greenwich Village, considered the "ground zero" of the AIDS epidemic, with one-third of its beds in 1986 filled by AIDS patients (Boynton 2013), is now luxury housing.

Rather than fight this transition through radical political activism as groups like ACT UP did, mainstream, homonormative gay political activism had turned towards fundamentally conservative issues like gay marriage, and as Schulman (2012: 144) argues, "nothing is more desexualizing than marriage." Timothy Stewart-Winter (2015), the author of *Queer Clout*, asks, "Will even a fraction of the energy and money that have been poured into the marriage fight be available to transgender people, homeless teenagers, victims of job discrimination, lesbian and gay refugees and asylum seekers, isolated gay elderly or other vulnerable members of our community?" As one activist commented, "The drag queens who started Stonewall are no better off today, but they made the world safe for gay Republicans. It's a bitter pill to swallow, but the people who make change are not the people who benefit from it" (quoted in Schulman 2012: 115). As direct action was overtaken by assimilation, gay politics became gentrified and gay life privatized, with elite gay men organizing against queer youth who socialize on the streets (Schulman 2012).

The decline of the gayborhood

Yet, the mainstreaming of the gayborhood does not negate the need for safe queer space. The "longing for queer community persists" (Doan and Higgins

2011: 20). As Schulman (2012: 40) states, "if all gays could live safely and openly in their communities of origin, and if government policies had been oriented towards protecting poor neighborhoods by rehabbing without displacement, then gentrification by white gay men would have been both unnecessary and impossible." The narrative of the gentrified gayborhood erases the history of struggle that formed these neighborhoods. Historically, the disappearance of gay enclaves is nothing new. Gay residential and commercial space has long been displaced when more desirable uses for the space are found. The transgender community in Istanbul is being targeted by large scale, state-sponsored gentrification; expulsion of trans people is state policy (TransX Istanbul 2014; see also Carter 2015). Papadopoulos (2017) argues that at least one of the goals of large scale urban renewal in Chicago was the expulsion of the gay enclave on the Gold Coast, pushing gays farther north to what would become known as Boystown, an area with a large rent gap. In the neoliberal city, the gay enclave is useful only insofar as it attracts capital, but that success is threatened by speculative national, and even international, circuits of capital which then serve to displace local gay entrepreneurs (Papadopoulos 2017). The consumption of these queer spaces by straight people "cannot be assumed to denote acceptance, or even tolerance" (quoted in Ghaziani 2014: 26).

The newfound attractiveness of the gayborhood to real estate developers and straight people, while beneficial to the few who can profit from it, displaces the very communities they are meant to celebrate. Such is the case with gay venues in London, where at least 12 gay venues, many in the iconic neighborhood of Soho, have faced closure since 2008 (Margolis 2015). One, the Joiners Arms, is to be turned into luxury flats. While the global recession has certainly had an impact, the gay users of these spaces feel that they are being targeted by governmental apathy combined with real estate forces to destroy London's gay scene. One protestor commented, "I think it's a very British, stiff upper lip, conservative way of making a specific attack on the community" (quoted in Margolis 2015). The experience is hardly unique to London. In San Francisco, gay spaces have been facing pressure from the tech boom, with a surge in the number of closures since 2012, including the long-lived Stud, opened in 1966, which saw a 300 percent increase in rent. "Stud has always fostered a radical queer community, from bikers and miscreants to working-class drag artists and performers like Janis Joplin and Etta James" (Terrell 2016).

The last lesbian bar in San Francisco, the Lexington Club, announced its closing in 2014, a reflection, Gieseking (2014) argues, of the fact that, "not only lesbians but *all* women bear the greater brunt of gentrification today, alongside people of color and the poor" as well as transgender people, all of whom are more likely to experience gentrification as the gentrified rather than as gentrifier. While gay enclaves may be under threat by gentrification in cities across the Global North, too little recognized is the fact that gender is an exacerbating factor.

Gieseking's (2013: 189) research with lesbians in the Park Slope neighborhood of Brooklyn shows just "how *unsafe* the economic land grab of gentrification is ... [G]entrification is clearly not a sustainable tactic of these women's resilience of resistance to homophobia and sexism because it eventually moves them out of those spaces they increased in value as well" (emphasis in original). LGBTQ people are erased from physical space and from planning documents. Doan and Higgins (2011) find that the gay enclave in Atlanta's Midtown neighborhood goes unmentioned as such in planning documents, as is true in many other cities (Chicago is an exception, see Stewart-Winter 2016). Their erasure from planning documents is mirrored by the gentrification of Midtown, which is displacing the gay population and becoming less tolerant of LGBTQ people and businesses. This displacement increases the vulnerability of LGBTQ people and restricts the ability of the community to organize and elect LGBTQ-friendly officials (Doan and Higgins 2011).

As Podmore (2013) comments, the significant amount of attention paid to the role of gay villages in urban regenerations has overshadowed the continued enforcement of heteronormativity through local planning, policy and law. Indeed, Podmore (2013) argues, the disdain for the gay village can be a metronormative and privileged position, often expressed by those who have both experienced the benefit of the gayborhood and have the luxury of moving beyond it. The gayborhood is being "de-queered," increasingly understood as "an artifact of historic activism" that is no longer necessary (Papadopoulos 2017). Ghaziani (2014) finds that it is, ironically, living in the gayborhood that enables gays and lesbians to believe that their sexual orientation is irrelevant. The normalization of the gayborhood allows some in a position of privilege to forget that housing and employment discrimination are real. A 2011 study found that same-sex couples were 25 percent more likely to be rejected by landlords seeking tenants, a rate of discrimination that disappears in gay neighborhoods (Ghaziani 2014).

Displacement is not the only danger. Some question whether the link between gentrification and gay neighborhoods is increasing prejudice and violence against LGBTQ people in gentrifying neighborhoods. Monroe (2013) places the death of Islan Nettles, a young trans woman who was murdered in Harlem by a man who had been teased for flirting with her, in the larger context of the gentrification of the neighborhood. The resentment toward white queer people moving to Harlem has been expressed more openly towards black queer people in the form of homophobic and transphobic slurs and attacks, a (misplaced) expression of the community's frustration with gentrification. A trans woman in Harlem was told by her neighbor that "she hated seeing the sight of me and my partner move in. She said to me, 'See what you bringing up in here,' referring to my girlfriend being white, 'and she'll be bringing more of her kind'" (quoted in Monroe 2013). The association of gay people with gentrification allows for a mis-identification of the causes of gentrification, ignoring structural capitalist forces seeking profit and the history of discrimination that made the gayborhood necessary.

Gender is an important element in these acts of aggression. Much of the violence experienced by gay and transgender people has to do not with sexuality, but with their presentation of gender, with effeminate men and masculine women the targets of aggression, and transgender people, especially trans women, at highest risk. Namaste (1996) argues that in separating gender from sexuality, the issue of gender has been foreclosed in gay male activism. Instead, activism needs to recognize the central role of gender in these aggressions, leading Namaste (1996) to call for organizing around "gender-bashing" rather than "queerbashing" as a way to establish a coalition among transgendered people, lesbians, bisexuals and gay men, recognizing that sexual and gender liberations are mutually bound.

Of course, not all queer people are equally at risk. Acceptance or exclusion is not monolithic but locally negotiated through race, gender, and class identities. So, for example, in South Africa, as elsewhere, white gay men can negotiate all manner of homophobia merely because in a racialized and classed society being "white" and broadly of the same "class" – the "us" – overrides sexuality as a major marker of difference (Visser 2013). Following the shooting at the Pulse nightclub in Orlando in 2016, media coverage highlighted the vulnerability to crime of the LGBTQ community. But in fact, those who are targeted are overwhelmingly people of color, prompting Halberstam (2016) to argue, "As middle class white LGBT people celebrate their access to normative social forms and agree to pay the price for such acceptance by consenting to new forms of violent exclusion, they/we cannot simultaneously claim to be the most vulnerable of the vulnerable."

Queer theory offers us alternative ways to think about space, community, and justice to counter the narrative of the gayborhood as a driver of gentrification and develop forms of resistance. Indeed, this is what LGBTQ activism was supposed to be about in the first place (Hanhardt 2013). Finding common cause with other oppressed people had been one of the ways gay political organizing became mainstream. The black civil rights movement provided a model for gay activism and also an opportunity to reject traditional political structures that were corrupt and unfair (Stewart-Winter 2016). Gay activists working with African-American and Latinx activists to protest police brutality in the Chicago neighborhood of Lincoln Park in the 1970s were at the forefront of contesting the gentrification of that neighborhood (Stewart-Winter 2016). We can see again today the potential for intersectional alliances around race and sexuality with the emergence of the Black Lives Matter movement, many of whose most prominent spokespeople are queer/women.

Queer futurity

Queering the neighborhood means the complete rejection of gentrification since gentrification relies on a competition over space that stigmatizes and displaces those who could broadly be categorized as queer. Muñoz (2009: 1) describes queerness as "essentially about the rejection of a here and now and an insistence on potentiality or concrete possibility for another world." Queerness is not yet here; it is a longing that propels us forward, employing hope as a critical methodology and celebrating collectivity (Muñoz 2009). So the political project of queerness goes far beyond sexuality, neither heteronormative nor reducible to gay or lesbian (Hanhardt 2013: 218). It should be viewed, Isoke (2014: 357) argues, as an "unwillingness, or perhaps even a glorious failure, to conform" either to the dictates of white normativity or the politics of respectability and in so doing forges a politics that celebrates the multiplicity and heterogeneity of identities. For Manalansan (2015: 567), "[q]ueer is also about the productive possibilities of people who are left out, displaced, disposed because of their position within the landscapes of the normal." As such, fighting gentrification can be framed as a queer issue. The queer activist organization Queer to the Left, based in Chicago argued, "'Queer' also embraces such struggles as that against gentrification, which is driven by an oppressive racist logic, as part of the same larger fight for social and economic justice" (quoted in Brown-Saracino 2009: 204).

So thinking about queer space requires we move beyond the homonormative assumptions about the gayborhood. Martinez (2015: 179) argues that scholars may be ringing the death knell of the gayborhood too soon, "bemoaning the death of queer communities in part because they are altogether missing the vibrancy of queer immigrant communities." She offers the Jackson Heights neighborhood of Queens in New York City as an example of a thriving, multicultural queer cosmopolis. Though largely unrecognized as such, it has been a queer space since the 1940s. It is home to LGBTQ elders as well as an incredibly diverse immigrant community (Colombian, Ecuadorian, Peruvian, Mexican, Puerto Rican, Dominican, Indian, Bangladeshi, Chinese, Filipino, Korean, Pakistani, and Nepalese), a safe haven for LGBTQ people who are not easily assimilated: interracial queer families, gender non-conforming people, LGBTQ seniors, and LGBTQ new immigrants and non-English speakers. The viability of Jackson Heights as a queer neighborhood is a result of this diversity, with the gay community, the Latinx community, and the Asian-American community voting as a progressive block (Martinez 2015). While Jackson Heights is not completely safe from the real estate pressures ravaging the rest of New York City, the traditional narrative of gay gentrification is complicated here by the diversity and intersectionality of its residents. Unlike in many other gayborhoods in Western cities, the LGBTQ population of Jackson Heights is growing, not shrinking (Martinez 2015).

Queering the neighborhood also means reconceptualizing to whom the neighborhood belongs. Gentrification has the effect of claiming space only for those who can afford to consume it. Ownership of real estate confers membership in the neighborhood. In that context, it is impossible for many queer people to "belong." Instead, activist collectives like FIERCE (Fabulous Independent Educated Radicals for Community Empowerment) in New York City have sought to claim space for those who actually use it, in this case the queer youth of color from all over the region who claim Greenwich Village as their own. They do this by emphasizing both the use value of the space for queer youth, but also the value that queer youth produce for the neighborhood (Hanhardt 2013). The presence of queer youth of color on the piers may be one of the few things helping to tamp down rampant development, thereby helping to preserve whatever affordability remains for renters and homeowners threatened with skyrocketing costs. But the stratification of neighborhoods made visible by the presence of queer youth of color in hyper-gentrified spaces exposes the uneven development of the city and the fact that "private property value requires poverty" (Hanhardt 2013: 212). A FIERCE action card succinctly explains the dynamic, "WHO PAY$ FOR YOUR QUALITY OF LIFE? Queer youth of color, Trannies, sex workers and the homeless are being cleaned out of the west village [sic]. But we are NOT trash" (quoted in Hanhardt 2013: 207). FIERCE has been successful in bringing the issues of queer youth into the political conversation, but gentrification continues to threaten this success. In the face of a new development plan for Hudson River Park, FIERCE ended its campaign for a physical safe space in the community, refocusing its efforts to city and nationwide change on the issue of policing (Goh 2015). As gentrification has threatened the loss of safe spaces in specific neighborhoods, queering the neighborhood has meant going beyond the geographic limits of one neighborhood, recognizing that a small number of safe spaces is not enough. Scaling up is a strategy to tackle the more structural issues that go far beyond the gayborhood. The story of queer activism is one of incomplete and uneven persistence, an alternative urban worldmaking that lays bare the strictures of neoliberalism and creates urban futures that deploy messes, limits and illegibilities to contest the narrative of unhampered progress that gentrification proponents sell (Manalansan 2015).

The people now most in danger of displacement from the gayborhood are also those paving the way towards queer futurity, rethinking notions of identity, family, and belonging in ways that are empowering and expansive. Andersson (2015: 280) recounts a scene from the iconic film about black drag culture, *Paris Is Burning*, in which queer kids describe their relationship, regardless of gender, as that of sisters, as "a queer mode of sociality beyond the mimicking of the straight nuclear family." This family is, of course, born out of the rejection from traditional family structures, but it is in these fissures that we can start to imagine alternative ways of being. As Andersson (2015: 280) argues:

> In the era of same-sex marriage-endorsed by Bloomberg [former mayor of New York City] ... with reference to the business case ('freedom attracts talent')- the celebration of such alternative modes of family and community are more important than ever because their vantage points,

which denaturalize kinship, are also uniquely well-placed to denaturalize the connection between property and propriety.

Truly queering community means rejecting many of the "common sense" narratives of what constitutes a "good" neighborhood and of conceptions of risk and value.

Gieseking (2013: 179) makes the case for rethinking the LGBTQ neighborhood as something beyond the normative paradigm of "property ownership as success" in order to recognize the experiences of women, working class people, and people of color. Park Slope, Brooklyn is an example of a neighborhood that has maintained its image as a lesbian enclave despite the lack of a spatial concentration of lesbians due to their displacement from gentrification (see Rothenberg 1995). Yet it has maintained its hold on the imagination as a lesbian enclave, understood as a lesbian space simply because people think that it is. In this way, Gieseking (2013) argues, lesbians and queer women have queered the meaning of the neighborhood by understanding and enacting it as fleeting and fragmented, intangible and imagined rather than fixed. This queering affords "the malleability to spatialize their sense of community, imagined and otherwise, and to do so to promote social and spatial justice" (Gieseking 2013: 196), making room for more and different types of scenes, especially with regards to race and class. These possibilities are always contextualized, so they will not look or be enacted the same way in different places.

Visser makes a broadly similar point, arguing that the idea of the gayborhood is a Western construct that does little to reflect the experience of those in cities in the Global South. Gay spaces in the form of consolidated physical spaces, he argues, are not a necessary outcome of lived gay identities. Rather, sexuality is often "lived through" networks of people, not in fixed neighborhood sized spaces (Visser 2013: 271). Thus, we need to recognize and celebrate the fluidity of sexual identity and spaces in which it is or might be expressed (Visser 2013).

The Black Lives Matter movement presents a vision of community and organizing that is queer, recognizing the intersectionality of identity and offering an alternative vision to things as they are. Activism around police brutality brings together queer/ people of color as the civil rights and gay rights movements did in the 1960s (Stewart-Winter 2016) and offers an alternative understanding of community that celebrates communities of care, a direct assault on the assumptions that underlie gentrification. Among the movement's guiding principles are commitments to building a "queer affirming network," "doing the work required to dismantle cis-gender privilege" and "disrupting the Western-prescribed nuclear family structure requirement by supporting each other as extended families and 'villages' that collectively care for one another" (http://blacklivesmatter.com/guiding-principles/). Queering the neighborhood in this way would fundamentally rework the way gender is enacted in the urban landscape, and I would argue, make gentrification impossible.

Conclusion

Queer spaces provide some of the best opportunities for challenging gender expectations and rethinking how gender is enacted on the urban landscape. The gentrification of queer spaces and gay spaces as gentrified enclaves foreclose these possibilities, reinforcing (white) male privilege and marginalizing those who do not conform to dominant norms of gender and respectability. Even in the most overtly gay spaces, there is very little gender queerness (Doan 2007). The demise of the gayborhood is dangerous for those who cannot perform the homonormativity that has become mainstream in many Western cities. Gentrification, in the end, is not different in gay neighborhoods. It accomplishes the displacement of what came before in order to reinforce a masculinist neoliberal vision of the urban. Whether it is through the displacement of lesbian enclaves, the police harassment of queer youth of color, or the stigmatization of transgender people visible in urban space, the gentrification of queer spaces has resulted in the re-gendering of these spaces as masculine, a direct refutation of the radical queer politics that birthed the gay rights movement.

As Schulman (2012) argues, not just the gayborhood but the gay imagination itself has been gentrified, this indeed being a necessary precursor to the accomplishment of a gentrified gayborhood. The historical struggles for gay rights, the trauma of the AIDS era, and lesbian, working class, and trans struggles have been gentrified out of the story of the gayborhood. "It's hard to have a collective memory when so many who were 'there' are not 'here' to say what happened" (Schulman 2012: 135). But the gentrification of gay neighborhoods that results in the displacement of even the most established of gay institutions makes it clear that the tolerance for these gay spaces is not mainstream, but provisional and fleeting.

Gentrification threatens the ability of the gayborhood as we know it to exist at all. But perhaps this will allow for the opportunity to imagine and create genuinely queer spaces in which gender is fundamentally reimagined, thereby undermining one of the fundamental ways in which power is maintained (Doan 2007). The displacement of queer spaces by gentrification and the recognition that in a neoliberal economy these enclaves will always be under threat can lead to a respatialization of our understanding of neighborhood and community that queers much more space than the historical gayborhood. Following Nash and Gorman-Murray (2014), it is not only queer spaces that must accommodate gender and sexual diversity; all spaces need to be inclusive of broad differences. Gentrification does not allow for this.

7   Conclusion

At a time when the UN's New Urban Agenda aims to:

> achieve gender equality and empower all women and girls, ensuring women's full and effective participation and equal rights in all fields and in leadership at all levels of decision-making, and by ensuring decent work and equal pay for equal work, or work of equal value for all women, as well as preventing and eliminating all forms of discrimination, violence and harassment against women and girls in private and public spaces (quoted in Moser 2016),

the global urban strategy of gentrification moves us in the opposite direction. Rather than serve as an emancipatory opportunity, gentrification has served to solidify class-based divisions, which are invariably gendered. A focus on the gendered inequalities of gentrification does not undermine class analysis, but rather serves to interrogate the dual oppressions of capitalism and patriarchy (McDowell 1983) as well as how these are further intensified and mediated through race, age, sexuality and ability. The previous chapters have illustrated the ways in which gender is implicated in the production of gentrified urban landscapes. Its meaning and uses are socially constructed, with gender employed to different ends depending on the context, in service to the interests of both capital and patriarchy. Gender may be employed as a strategy to claim space, to market new neighborhoods, and to stigmatize certain populations, but in most cases, this results in an increased burden on and marginalization of working class women and others who do not fit idealized gender norms. Inclusion in the labor and housing markets has not meant liberation for women, but rather the adjustment of women to masculinist market norms. As Wichterich (2015) argues, the neoliberal empowerment of women is instrumental to flexible, unregulated and precarious labor markets and thus to the maximization of productivity and growth. Exploitation and discrimination are in complex interplay with recognition and empowerment. Class, gender, race, and ethnicity have been used as wedges to politically isolate different constituencies within the urban core. Accomplishing the reinforcement of these differences has been a strategy to preclude alliances across boundaries. I suggest gender as one fruitful avenue through which class and racial divides can be crossed to build an alternative vision of urban redevelopment that contests gentrification as the only path to revitalization.

Rose (1984) envisioned this potential in her work on marginal gentrifiers. While much of the visible change wrought by gentrification reflects the influx of wealth necessary to accomplish gentrification, individual people we might call gentrifiers are not necessarily as wealthy as their working class neighbors might assume. By marginal gentrifiers, Rose (1984) was referring to those who chose inner city locations by necessity, to balance work and family responsibilities, for its relative cheapness, or who became reluctant buyers, purchasing condos in order to stay in buildings where they had been long-term renters. While these people hold relative privilege compared to those who cannot, under any circumstances, afford to stay in gentrifying neighborhoods, they are hardly the global oligarchs driving speculative investments in real estate in global cities. While these people are not the drivers of gentrification, their relative privilege differentiates them in terms of class position from their more working class neighbors. But, Rose (1984) argued, because of the restructuring of white collar work, the roll-back of public sector wages, and reduced job security, these people are significantly less secure than their counterparts a generation before (those in academia are one such example, as in the Fight for 15, discussed in Chapter 3). As true as this was in 1984, it is even more notable now, with the financial crisis of 2008 and beyond further accomplishing the proletarianization of white collar jobs and structural insecurity across the labor market. Thus, the struggles of marginal gentrifiers may be broadly similar to their neighbors, though they are rarely understood as such. This is especially true among certain groups such as single mothers, who not only rely on one income, but are also unlikely to see their incomes rise at the same rate as their male counterparts. Even those with higher incomes will have to spend a significant portion of it on child care.

To focus then on cultural or lifestyle choices of gentrifiers, as many popular representations of gentrification do (for a critique see Slater 2006), ignores the questions of housing affordability, economic restructuring, and the lack of an infrastructure of care that are driving people to make the locational decisions that they do. Rather, Rose (1984: 68) argued, it would be useful to focus on the needs that marginal gentrifiers share with their working class neighbors and those they have displaced. If we do not, "we are preventing any recognition of the possibility of forming alliances" between marginal gentrifiers and those likely to be displaced, whereas finding common cause may allow for the reshaping both of the physical fabric of the neighborhood and social networks within them. These alliances may support ways to rework traditional patriarchal divisions of labor and offer collective strategies to contest the neoliberal ideology of individualism. She offers the "'sphere of everyday life' and the reproductive work carried on therein" as the logical starting points of this alternative political practice (Rose 1984: 68).

Alliances around shared gender and other oppressions offer prefigurative ways of living and working. "Prefiguration is an effort to bring desired futures actively into being in the present. It promotes valuing the quality of everyday experience, and the processes of achieving change, as central to the doing of politics" (Siltanen et al. 2015: 263). This may be in the mundane, everyday decisions we make about how to speak and listen to people, acknowledging them in such a way as to build the bonds necessary to sustain action for change (Siltanen et al. 2015: 265). I see this call for prefigurative urban geographies as part of the project to reclaim care as radical practice, creating new collectivities and publics, for caring represents a politics of resistance to the dominant neoliberal constructs of value (Cheng 2016) that have been so essential to accomplishing gentrification. I would like to propose an ethic of care as a way to both contest gentrification and accomplish socially just urban development. This requires us to see our self interest in the things that we share (Curran 2016). Care is essential to both social reproduction and sustainable communities and requires a redefinition of work and value that is fair and gender-just (Wichterich 2015). A politics of care would make gentrification impossible.

Other ways of knowing our place

Contesting gentrification requires the production of other ways of knowing our place, and organizing around gendered issues can be one way to enact these alternative claims to space. As Rahder and McLean (2013: 161) highlight in their work on how immigrant women make community in the gentrifying city, given that "women play a vital role in creating the social networks that allow our communities to take root, a feminist environmental justice framework becomes key to establishing other ways of knowing our place." While "knowing your place denotes social exclusion and the subtle or not-so-subtle ways that marginalized groups are blocked from using needed public spaces," an exclusion that is often gendered given the history of women's more limited access to public space, "*other ways of knowing* your place" involve developing the social commitments and spatial attachments necessary for sustainable communities (Rahder and McLean 2013: 145, emphasis in original). Activism and the creation of supportive communities can change what we think we know about how cities work and who has the right to them.

One such example I highlighted in Chapter 2 is that of the Focus E15 Mothers, whose organizing helped the mothers and other activists rethink their place in the city and their ability to affect change. Their creative and joyful campaign has given rise to the high-profile reputation of the Focus E15 campaigners as inspirational young women who do not "know their place" (Watt 2016: 297). They have created a new way of knowing their place, first by engaging in activism to claim their right to stay in London in the first place, exhibiting a power they didn't realize they had, but then by creating a movement that is generative, reaching out across the city and across race and class, to inspire similar movements elsewhere. In this way, it expands the constituency of the movement to anyone at risk in London's speculative housing market. This broad reach means the campaign does not "represent" any singular identity but instead inveigles class, place, gender, motherhood, generation, and race (Watt 2016: 316).

This "becoming together" (Watt 2016) illustrates Focus E15's collective project of making a claim to space, through their weekly street stall, their occupation of closed housing estates, their parties in show apartments, and now their new office space, that allows for new ways of knowing their place which help to then create alternative realities which contest gentrification and social cleansing as inevitable.

Another potential strategy is that of commoning. Commons are generally understood as collective ownership of land or other resources held in a communal manner, sometimes in opposition to private property, and can be a way to challenge the traditional binaries of individual and society, public and private, state and market (Noterman 2016). In an era of rampant real estate speculation and gentrification coupled with austerity, commoning has become a strategy through which people meet concrete needs, and in so doing confront dynamics of power (Bresnihan and Byrne 2015). Bresnihan and Byrne (2015) cite the inability to access space for anything that is not profit-generating in the high rent environment of Dublin as the motivation for participating in common spaces. Here, the commons is an attempt to contest the enclosure of urban life, as only those uses seen as valuable by the market have any place in the city. The commons are a way of "materializing an alternative" (Bresnihan and Byrne 2015: 48) in which spaces belong, first and foremost, to those who participate in and make use of them. Commoning as a process, therefore, requires new social contracts, with rights, duties and rules on how to share and care for common resources (Wichterich 2015). While often temporary, examples of commoning teach us new ways to be in the city. Examples in Dublin cited by Bresnihan and Byrne (2015) include meeting spaces, internet access, bicycle maintenance, educational workshops, and a food cooperative. They also include examples of citizen initiatives to open spaces when city administrations plan to close them down, as with the occupation of La Casita as a volunteer-run neighborhood library in Pilsen discussed in Chapter 4. Despite the obvious problems of simply finding and holding space in the gentrified city, alternative ways of being emerge. For Bresnihan and Byrne (2015: 46) this was evident any time they were involved in the preparation of large meals in common space:

> Each time we helped prepare one of these meals, there was an apparent lack (of time, space, knowledge, resources), but through the socialization of production, the sharing of space, the sharing of implements and the sharing of knowledge, the situation was always transformed and problems overcome.

Of course, the working class is very experienced in the art of making do. This collectivity needs to be politicized as a way to reclaim and remake urban space.

Becoming mothers

Returning again to the Focus E15 Mothers as a point of departure, I'd like to expand on a way of becoming that Paul Watt (2016) highlights in his article on the successes of Focus E15. In addition to the mothers becoming activists, activists and others who attend the weekly stall or other events, as part of their participation, become mothers, caring for Focus E15 mothers' children when the mothers were at the microphone or otherwise engaged, thus blurring "fixed" social identities and boundaries (Watt 2016: 317). Mothering here is not biological, but rather relates to the *work* of parenting, following Ruddick (1992:185 emphasis in original), in which both men and women engage. Ruddick makes the case for mothering as a gender-inclusive and therefore genderless activity, in which a person is a mother if they act upon a social commitment to nurture, protect, and train children. Using the term mother, rather than parent, acknowledges that mothering has been primarily the responsibility of women, that this history has consequences, and that we need to construct a material alternative (Ruddick 1992).

In the context of resisting gentrification, I would like to argue that we must all become mothers, not just of (our own) children, but more broadly, engaging in care work that helps to materialize the alternative ways of knowing our place that we would like to build and live.

Conclusion

In making the argument for a greater focus on the role of gender in understanding the effects of gentrification, I am not arguing that we do so in place of interrogating class and race, or any other way of "othering" that has helped to solidify difference in the urban landscape. Rather, following Haraway (2013), I am interested in "staying with the knots, staying with the trouble" in order to understand more fully. Gender is but one of the knots of gentrification. The focus of resources on gentrifying areas of the city goes hand in hand with the neglect of other working class neighborhoods, a recreation and reinforcement of the landscape of uneven development that allows for the rent gap. This is true both within individual cities and across cities, as too many cities compete for the few who are capable of affording luxury developments in the gentrified city. This competition is masculinist, accomplishing the regendering of the city, putting women and those who are in any way othered at further disadvantage in housing and labor markets, increasing the burden of social reproduction, undermining democracy, and making cities less safe for their long-term residents. With gender still very much at the core of how the negative effects of gentrification are experienced, gender analysis is essential as an organizing tool to contest the vision of the city that gentrification has wrought.

References

Addams, J. (1913) *Women and Public Housekeeping*. http://search.lib.virginia.edu/cata log/uva-lib:770642/view#openLayer/uva-lib:956744/4040.4724718352/2097/2/1/0 Accessed July 28, 2016.

Aitken, S.C. (1998). *Family Fantasies and Community Space*. New Brunswick, NJ: Rutgers University Press.

Andersson, J. (2015) 'Wilding' in the West Village: queer space, racism and Jane Jacobs hagiography. *International Journal of Urban and Regional Research*, 39, pp. 265–283, doi:10.1111/1468-2427.12188

Anguelovski, I. (2013). From environmental trauma to safe haven: place attachment and place remaking in three marginalized neighborhoods of Barcelona, Boston, and Havana. *City & Community*. 12(3), pp. 211–237.

Anguelovski, I. (2014) *Neighborhood as Refuge: Community Reconstruction, Place-Remaking, and Environmental Justice in the city*. Cambridge: MIT Press.

Anguelovski, I. (2015) Alternative food provision conflicts in cities: contesting food privilege, injustice, and whiteness in Jamaica Plain, Boston. *Geoforum* 58, pp. 184–194.

Arnold, G. and Slusser, M. (2015) Silencing women's voices: nuisance property laws and battered women. *Law and Social Inquiry* 40(4), 908–936.

Atkinson, R. (2000) Measuring gentrification and displacement in Greater London. *Urban Studies* 37, pp. 149–165.

Badger, E. (2013) The insane, true costs of raising a family in America's major metros. *CityLab*. July 12. http://www.citylab.com/housing/2013/07/insane-true-costs-raising-family-americas-major-metros/6172/ Accessed July 13, 2016.

Banzhaf, H.S. and McCormick, E., (2007). Moving beyond cleanup: identifying the crucibles of environmental gentrification. Andrew Young School of Policy Studies Research Paper Series, Working Paper 07-29, May, Georgia State University, Atlanta, GA.

Barry, U. (2014) Gender perspective on the economic crisis: Ireland in an EU context. *Gender, Sexuality & Feminism*. 1(2), pp. 82–103.

Beattie, J. (2014) Pride of Ireland: five women who turned their crumbling estate into top-notch building. *Irish Mirror*. March 9. http://www.irishmirror.ie/female/pride-ir eland-meet-five-women-3197887 Accessed November 8, 2016.

Beauregard, R.A. (1986) The chaos and complexity of gentrification. In Smith, N. and Williams, P., eds. *Gentrification and the City*. Boston: Allen and Unwin, pp. 35–55.

Bell, D. and Valentine, G., eds. (1995) *Mapping Desire: Geographies of Sexualities*. London and New York: Routledge.

Bellafante, G. (2016) Poor, transgender and dressed for arrest. *New York Times.* October 2, p. 28.

Bennett, L., Smith, J., and Wright, P.A. (2006) *Where Are Poor People to Live? Transforming Public Housing Communities.* Armonk, NY: M.E. Sharpe.

Betancur, J. and Smith, J. (2016) *Claiming Neighborhood: New Ways of Understanding Neighborhood Change.* Urban, Chicago and Springfield: University of Illinois Press.

Binnie, J. (1995) Trading places: consumption, sexuality and the production of queer space. In Bell, D. and Valentine, G., eds. *Mapping Desire: Geographies of Sexualities.* London and New York: Routledge, pp. 182–199.

Binnie, J. and Skeggs, B. (2004) Cosmopolitan knowledge and the production and consumption of sexualized space. *The Sociological Review* 52(1), pp. 39–61.

Blanco, I. and Subirats, J. (2008) Social exclusion, area effects and metropolitan governance: A comparative analysis of five large Spanish cities . *Urban Research and Practice* 1, pp. 130–148.

Blau, F. and Kahn, L. (2016) The gender wage gap: Extent, trends, and explanations. IZA Discussion Paper 9656. January. http://ftp.iza.org/dp9656.pdf Accessed May 23, 2016.

Bondi, L. (1991a) Gender divisions and gentrification: a critique. *Transactions of the Institute of British Geographers* 16, pp. 190–198.

Bondi, L. (1991b) Women, gender relations and the inner city. In Keith, M., and Rogers, A., eds. *Hollow Promises: Rhetoric and Reality in the Inner City.* London: Mansell, pp. 110–126.

Bondi, L. (1999) Gender, class and gentrification: enriching the debate. *Environment and Planning D: Society and Space* 17, pp. 261–282.

BOOM: The Sound of Eviction. (2002) A Whispered Media Production. Directed by F. Cavanaugh, A.M. Liiv, J. Taylor, and A. Wood. Boterman, W.R. (2013) Dealing with diversity: middle-class family households and the issue of 'black' and 'white' schools in Amsterdam. *Urban Studies* 50(6), pp. 1130–1147.

Boterman, W.R. and Karsten, L. (2014) On the spatial dimension of the gender division of paid work in two-parent families: the case of Amsterdam, the Netherlands. *Tijdschrift voor Economische en Social Geografie* 105(1), pp. 107–116.

Boterman, W.R. and Bridge, G. (2015) Gender, class and space in the field of parenthood: comparing middle-class fractions in Amsterdam and London. *Transactions of the Institute of British Geographers*, 40(2), pp. 249–261, doi:10.1111/tran.12073

Bowling, B. (1999) The rise and fall of New York murder: zero tolerance or crack's decline? *British Journal of Criminology* 39(4), pp. 531–554.

Boynton, A. (2013) Remembering St. Vincent's. *New Yorker.* May 16. http://www.new yorker.com/culture/culture-desk/remembering-st-vincents Accessed December 1, 2016.

Boyd, B. (2016) People of Stoneybatter prepare to be gentrified. *Irish Times.* February 3. http://www.irishtimes.com/opinion/brian-boyd-people-of-stoneybatter-prepare-to-be-gentrified-1.2519806 Accessed August 8, 2016.

Braedley, S. and Luxton, M. (2010) Competing philosophies: neoliberalism and challenges of everyday life, in Braedley, S. and Luxton, M., eds. *Neoliberalism and Everyday Life.* Montreal: McGill-Queens University Press.

Bresnihan, P. and Byrne, M. (2015) Escape into the city: everyday practices of commoning and the production of urban space in Dublin, *Antipode*, 47, pp. 36–54, doi:10.1111/anti.12105

Bridge, G., Butler, T. and Lees, L. (2012) *Mixed Communities: Gentrification by Stealth?*Bristol: Policy Press.

Brown, M. (2014) Gender and sexuality II: there goes the gayborhood? *Progress in Human Geography* 38(3), pp. 457–465, doi:10.1177/0309132513484215

Brown, T.M. (2011) *Raising Brooklyn: Nannies, Childcare, and Caribbeans Creating Community.* New York and London: New York University Press.

Brown-Saracino, J. (2009) *A Neighborhood that Never Changes: Gentrification, Social Preservation, and the Search for Authenticity.* Chicago and London: University of Chicago Press.

Brown-Saracino, J. (2016), An agenda for the next decade of gentrification scholarship. *City & Community*, 15, pp. 220–225, doi:10.1111/cico.12187

Brownill, S. (2000) Regen(d)eration: women and urban policy in Britain. In Darke, J., Ledwith, S. and Woods, R., eds. *Women and the City: Visibility and Voice in Urban Space.* Basingstoke and New York: Palgrave, pp. 11–40.

Brownlow, A. (2009) Keeping up appearances: profiting from patriarchy in the nation's 'safest city.' *Urban Studies*, 46(8), pp. 1680–1701.

Budig, M., Misra, J., and Boeckmann, I. (2012) The motherhood penalty in cross-national perspective: the importance of work-family policies and cultural attitudes. *Social Politics* 19(2), pp. 163–193.

Buffel, T. and Phillipson, C. (2016) Can global cities be 'age-friendly' cities? Urban development and ageing populations. *Cities* 55, pp. 94–100.

Burke, C.W. with the Better Government Association. (2014) The rise and fall of Juan Rangel, the Patrón of Chicago's UNO Charter Schools . *Chicago Magazine.* February. http://www.chicagomag.com/Chicago-Magazine/February-2014/uno-juan-rangel/ Accessed July 27, 2016.

Butler, T. and Robson, G. (2003) Plotting the middle classes: gentrification and circuits of education in London. *Housing Studies* 18(1), pp. 5–28.

Byrne, M. (2016) Entrepreneurial urbanism after the crisis: Ireland's "bad bank" and the redevelopment of Dublin Docklands. *Antipode*, doi:10.1111/anti.12231

Cahill, C. (2007) Negotiating grit and glamour: young women of color and the gentrification of the Lower East Side. *City & Society*, 19, pp. 202–231, doi:10.1525/city.2007.19.2.202

Cahill, C. (2006) "At risk"? The fed up honeys re-present the gentrification of the Lower East Side. *Women's Studies Quarterly* 34 (1&2), pp. 334–363.

Carpenter, J. and Lees, L. (1995) Gentrification in New York, London and Paris: an international comparison. *International Journal of Urban and Regional Research* 19, pp. 286–303.

Carter, W. (2015). Suffering and loss in Istanbul's transgender slum. *Middle East Eye* July 28. http://www.middleeasteye.net/in-depth/features/silence-suffering-and-dispossession-istanbul-s-transgender-slum-1343803757 Accessed December 9, 2016.

Cassidy, J. (2015) College calculus: what's the real value of higher education? *New Yorker.* September 7. http://www.newyorker.com/magazine/2015/09/07/college-calculus Accessed May 6, 2016.

Castells, M. (1983) *The City and the Grassroots.* Berkeley, CA: University of California Press.

Caulfield, J. (1989) 'Gentrification' and desire. *Canadian Review of Sociology*, 26(4), pp. 617–632.

Cele, S. (2015) Childhood in a neoliberal utopia: planning rhetoric and parental conceptions in contemporary Stockholm. *Geografiska Annaler: Series B, Human Geography*, 97, pp. 233–247, doi:10.1111/geob.12078

Center for Urban Research and Learning, Davis, J. L., and Merriman, D. F. (2007) One and a half decades of apartment loss and condominium growth: changes in Chicago's residential building stock. Center for Urban Research and Learning: Publications and Other Works. Paper 13. http://ecommons.luc.edu/curl_pubs/13

Centre for Urban Studies, ed. (1964) *London: Aspects of Change*. London: MacKibbon and Kee.

Chaban, M.V. (2016) With plan for Brooklyn hospital, neighbors may finally get their way. *New York Times* January 18. http://www.nytimes.com/2016/01/19/nyregion/with-plan-for-greenpoint-hospital-neighbors-may-finally-get-their-way.html?_r=0 Accessed July 29, 2016.

Chaskin, R. J. and Joseph, M.L. (2015) *Integrating the Inner City: The Promise and Perils of Mixed-Income Public Housing Transformation*. Chicago: The University of Chicago Press.

Chaskin, R., Joseph, M., Voelker, S. and Dworsky, A. (2012) Public housing transformation and resident relocation: comparing destinations and household characteristics in Chicago. *Cityscape: A Journal of Policy Development and Research* 14(1), pp. 183–214.

Checker, M., (2011). Wiped out by the "Greenwave": environmental gentrification and the paradoxical politics of urban sustainability. *City & Society*, 23(2), pp. 210–229.

Cheng, S. (2013) At the intersection of urban renewal and anti-trafficking projects: neoliberalism and a red-light district in Seoul, South Korea. *Scholar & Feminist Online* 11.1–11.2. http://sfonline.barnard.edu/gender-justice-and-neoliberal-transformations/at-the-intersection-of-urban-renewal-and-anti-trafficking-projects-neoliberalism-and-a-red-light-district-in-seoul-south-korea/ Accessed November 3, 2016.

Cheng, Y. (2016) Critical geographies of education beyond "value": moral sentiments, caring, and a politics for acting differently. *Antipode* 48, pp. 919–936, doi:10.1111/anti.12232

Cho, M. and Kim, J. (2016) Coupling urban regeneration with age-friendliness: neighborhood regeneration in Jangsu Village, Seoul. *Cities* 58, pp. 107–114.

Church, E. (2012) Toronto's deputy mayor faces backlash over disparaging downtown living. *The Globe and Mail*. July 12. http://www.theglobeandmail.com/news/toronto/torontos-deputy-mayor-faces-backlash-over-disparaging-downtown-living/article4411595/ Accessed July 13, 2016.

Clarke, P. (2015) Renaissance 2010 launched to create 100 new schools. *Catalyst Chicago*. May 22. http://catalyst-chicago.org/2015/05/renaissance-2010-launched-to-create-100-new-schools/ Accessed July 22, 2016.

Cohen, R. (2015) Baltimore can't rely on "Judge Judy" to protect renters. *Next City*. December 9, 2015. https://nextcity.org/daily/entry/baltimore-rent-court-report-released-evictions-reforming-rent-court Accessed January 21, 2016.

Coleman, R., Tombs, S., and Whyte, D. (2005) Capital, crime control and statecraft in the entrepreneurial city. *Urban Studies* 42(13), pp. 2511–2530.

Collins, A. (2004) Sexual dissidence, enterprise and assimilation. *Urban Studies* 41, pp. 1789–1806.

Cox, T. (2016) Can a little rebranding save struggling public schools in Chicago? *DNAInfo Chicago.* July 13. https://www.dnainfo.com/chicago/20160713/downtown/can-little-rebranding-save-struggling-public-schools-chicago Accessed July 22, 2016.

Cubanski, J., Cassillas, G., Damico, A. (2015) Poverty among seniors: an updated analysis of national and state level poverty rates under the official and supplemental poverty measures. Menlo Park, CA: Kaiser Family Foundation. June 10. http://kff.org/medicare/issue-brief/poverty-among-seniors-an-updated-analysis-of-national-and-state-level-poverty-rates-under-the-official-and-supplemental-poverty-measures/. Accessed January 28, 2016.

Curran, W. (2004) Gentrification and the changing nature of work: exploring the links in Williamsburg, Brooklyn. *Environment and Planning A.* 36(7), pp. 1243–1260.

Curran, W. (2007) 'From the frying pan to the oven': gentrification and the experience of industrial displacement in Williamsburg, Brooklyn. *Urban Studies* 44(8), pp. 1427–1440.

Curran, W. (2010). In defense of old industrial spaces: manufacturing, creativity, and innovation in Williamsburg, Brooklyn. *International Journal of Urban and Regional Research* 34(4), pp. 871–885.

Curran, W. (2014) Snow's not sexist, but the city's response is. *Daily Beast.* January 22. http://www.thedailybeast.com/witw/articles/2014/01/22/snow-s-not-sex ist-but-the-city-s-response-is.html Accessed July 13, 2016.

Curran, W. (2016) Gentrification is not inevitable: care and resistance. TEDx Talk. https://www.youtube.com/watch?v=yj1H8Sdc8Sw

Curran, W. and Breitbach, C. (2010) Notes on women in the global city: Chicago. *Gender, Place & Culture,* 17(3), pp. 393–399, doi:10.1080/09663691003737678

Curran, W. and Hague, E. (2006) Pilsen building inventory report. DePaul University Department of Geography. https://steans.depaul.edu/docs/PilsenInventory/Pilsen%20Report%20with%20graphs%202006.pdf Accessed December 7, 2016.

Curran, W. and Hague, E. (2013) Students and sit-ins at Whittier Elementary School. *AREA Chicago.* Winter/Spring, p. 55.

Curran, W., and Hamilton, T. (2012) Just green enough: contesting environmental gentrification in Greenpoint, Brooklyn. *Local Environment: The International Journal of Justice and Sustainability,* 17(9), pp. 1027–1042.

Curran, W. and Hanson, S. (2005) Getting globalized: city policy and industrial displacement in Williamsburg, Brooklyn. *Urban Geography* 26(6), pp. 461–482.

Cusk, R. (2001) *A Life's Work: On Becoming a Mother.* New York: Picador.

Dale, A. and Newman, L.L., (2009). Sustainable development for some: green urban development and affordability. *Local Environment: The International Journal of Justice and Sustainability,* 14(7), pp.669–681.

Dastrup, S., Ellen, I., Jefferson, A., Weselcouch, M., Schwartz, D. and Cuenca, K. (2015) The effects of neighborhood change on New York City Housing Authority residents. Prepared for the NYC Center for Economic Opportunity in partnership with the NYU Furman Center for Real Estate and Urban Policy. May 21. http://www.nyc.gov/html/ceo/downloads/pdf/nns_15.pdf. Accessed November 20, 2015.

Davidson, M. (2008) Spoiled mixture: where does state-led 'positive' gentrification end? *Urban Studies* 45(12), pp. 2385–2405.

Davidson, M. (2009) Displacement, space and dwelling: placing gentrification debate. *Ethics, Place & Environment: A Journal of Philosophy and Geography,* 12(2), pp. 219–234.

Davidson, M. (2012) The impossibility of social mixing. In Bridge, G., Butler, T. and Lees, L. *Mixed Communities: Gentrification by Stealth?*Bristol: Policy Press, pp. 233–250.

Davis, L.K. (2004) Reshaping Seoul: redevelopment, women and insurgent citizenship. PhD Dissertation. Baltimore, MA: Johns Hopkins University.

DeFilippis, J. (2004) *Unmasking Goliath: Community Control in the Face of Global Capital*. New York: Routledge.

de Koning, A. (2015) "This neighborhood deserves an espresso bar too": neoliberalism, racialization, and urban policy. *Antipode* 47:1203–1223, doi:10.1111/anti.12150

De la Torre, M., Gordon, M.F., Moore, P. and Cowhy, J. with Jagešić, S. and Huynh, M.H. (2015) *School Closings in Chicago: Understanding Families' Choices and Constraints for New School Enrollment*. University of Chicago Consortium on Chicago Schools Research. January. https://consortium.uchicago.edu/sites/default/files/publications/School%20Closings%20Report.pdf Accessed July 22, 2016.

DePillis, L. (2014) It's hard to build cities for kids. But do we really need them? *Washington Post*. August 19. https://www.washingtonpost.com/news/storyline/wp/2014/08/19/its-hard-to-build-cities-for-kids-but-do-they-really-need-them/ Accessed July 19, 2016.

DeSena, J. (2006) "What's a mother to do?" Gentrification, school selection, and the consequences for community cohesion. *The American Behavioral Scientist* 50(2), pp.241–257.

DeSena, J. (2009) Gentrification, schooling and social inequality. *Educational Research Quarterly* 33(1), pp. 60–74.

Desmond, M. (2016). *Evicted: Poverty and Profit in the American City*. New York: Crown Publishers.

Dillon, L. (2014) Race, waste and space: brownfield redevelopment and environmental justice at the Hunter's Point shipyard. *Antipode* 46(5), pp. 1205–1221, doi:10.1111/anti.12009

Ding, L., Hwang, J. and Divringi, W. (2015) Gentrification and residential mobility in Philadelphia. Federal Reserve Bank of Philadelphia Discussion papers. October. file:///C:/Users/winifred/Downloads/discussion-paper_gentrification-and-residentia l-mobility.pdf Accessed November 20, 2015.

DNAInfo Staff. (2014) Whittier playground officially opens on site of former fieldhouse. *DNAInfo*May 31. https://www.dnainfo.com/chicago/20140531/pilsen/whittier-playground-officially-on-site-of-former-fieldhouse Accessed July 27, 2016.

Doan, P. (2007) Queers in the American city: transgendered perceptions of urban space. *Gender, Place and Culture* 14(1), pp. 57–74.

Doan, P., ed. (2015) *Planning and LGBTQ Communities: The Need for Inclusive Queer Spaces*. New York and London: Routledge.

Doan, P. and Higgins, H. (2011) The demise of queer space? Resurgent gentrification and the assimilation of LGBT neighborhoods . *Journal of Planning Education and Research* 31(1), pp. 6–25.

Dooling, S., (2009). Ecological gentrification: a research agenda exploring justice in the city. *International Journal of Urban and Regional Research*, 33(3), pp. 621–639.

DPD (Chicago Department of Planning and Development) and CMAP (Chicago Metropolitan Agency for Planning) (2015). *Pilsen-Little Village Land Use Plan: Existing Conditions*. February. http://www.cmap.illinois.gov/programs-and-resources/lta/pilsen-little-village Accessed November 8, 2016.

The Economist. (2014) The glass ceiling index. *The Economist*, March 8, p. 70.

England, K. (1991) Gender relations and the spatial structure of the city. *Geoforum* 22 (2), pp. 135–147.

England, P. and Folbre, N. (1999). The cost of caring. *Annals of the American Academy of Political and Social Science*. 561(1), pp. 39–51.

Erbentraut, J. (2012) Wang's, Chicago gay bar, investigated over reports that it bans women during busiest hours. *Huffington Post*March 8. http://www.huffingtonpost.com/2012/03/08/wangs-chicago-gay-bar-ban_n_1327766.html Accessed August 24, 2016.

Faircloth, C., Hoffman, D. M., and Layne, L. L. (2013) *Parenting in Global Perspective: Negotiating Ideologies of Kinship, Self and Politics*. London: Routledge.

Fincher, R. (2004) Gender and life course in the narratives of Melbourne's high-rise housing developers. *Australian Geographical Studies* 42(3), pp. 325–338.

Flaherty, C. (2015). Fight for 15K. *Inside Higher Ed*. April 16. https://www.inside highered.com/news/2015/04/16/adjuncts-participate-national-day-action-living-wage Accessed November 29, 2016.

Florida, R. (2002). *The Rise of the Creative Class: And How It's Transforming Work, Leisure, Community and Everyday Life*. New York: Perseus Book Group.

Florida, R. (2016) The racial divide in the creative economy. *CityLab*. May 9. http://www.citylab.com/work/2016/05/creative-class-race-black-white-divide/481749/?utm_source=nl__link1_050916 Accessed May 18, 2016.

Focus E15. (2016a) Voices of Focus E15 mums nearly 3 years on. Focus E15. May 16. https://focuse15.org/2016/05/16/voices-of-focus-e15-mums-nearly-3-years-on/ Accessed September 27, 2016.

Focus E15. (2016b) More Focus mothers contact the campaign. Focus E15. May 17. https://focuse15.org/2016/05/17/more-focus-mothers-contact-the-campaign/ Accessed September 27, 2016.

Focus E15. (2016c) 3 years of resistance: how we did it. Focus E15. September 23. http s://focuse15.org/2016/09/23/3-years-of-resistance-how-we-did-it/ Accessed September 27, 2016.

Freeman, L. (2006) *There Goes the 'Hood: Views of Gentrification from the Ground Up* Philadelphia: Temple University Press.

Freeman, L. and Braconi, F. (2002) Gentrification and displacement. *The Urban Prospect: Housing, Planning and Economic Development in New York* 8(1), pp. 1–4.

Fullilove, M.T. (2004) *Root Shock: How Tearing Up City Neighborhoods Hurts America, and What We Can Do About It*. New York: Ballantine Books.

Fyfe, N. (2004) Zero tolerance, maximum surveillance? Deviance, difference and crime control in the late modern city. In Lees, L., ed. *The Emancipatory City? Paradoxes and Possibilities*. London: Sage, pp. 40–56.

Galcanova, L., and Sykorova, D. (2015) Socio-spatial aspects of ageing in an urban context" An example from three Czech Republic cities. *Aging & Society* 35, pp. 1200–1220.

Garber, J.A. (1995) Defining feminist community: Place, choice, and the urban politics of difference. In Garber, J.A. and Turner, R.S., eds. *Gender in Urban Research*. Thousand Oaks, CA: Sage, pp. 24–43.

Ghaziani, A. (2014) *There Goes the Gayborhood?* Princeton: Princeton University Press.

Gieseking, J.J. (2013) Queering the meaning of 'neighborhood': reinterpreting the lesbian-queer experience of Park Slope, Brooklyn, 1983–2008. In Taylor, Y. and

Addison, M., eds. *Queer Presences and Absences*. New York: Palgrave Macmillan, pp. 178–200.

Gieseking, J.J. (2014) On the closing of the last lesbian bar in San Francisco: what the demise of the Lex tells us about gentrification. *Huffington Post*. October 28. http://www.huffingtonpost.com/jen-jack-gieseking/on-the-closing-of-the-las_b_6057122.html Accessed August 29, 2016.

Glass, R. (1964) Introduction: aspects of change. In Centre for Urban Studies, ed. *London: Aspects of Change*. London: MacKibbon and Kee.

Goh, K. (2015) Place/out: planning for radical queer activism. In Doan, P., ed., *Planning and LGBTQ Communities: The Need for Inclusive Queer Spaces*. New York and London: Routledge, pp. 217–234.

Goldin, C. (2014) A grand gender convergence: its last chapter. *American Economic Review*. 104(4), pp. 1091–1119.

Goldin, C. and Dubner, S.J. (2016) The true story of the gender pay gap. *Freakonomics Radio*. January 7. http://freakonomics.com/podcast/the-true-story-of-the-gender-pay-gap-a-new-freakonomics-radio-podcast/. Accessed May 17, 2016.

Grossman, S. (2014) This map shows just how big the wage gap between men and women is. *Time*. March 6. http://time.com/14153/global-gender-pay-gap-map/ Accessed September 21, 2015.

Gulson, K.N. (2007) Repositioning schooling in Inner Sydney: urban renewal as education market and the 'absent presence' of the 'middle classes'. *Urban Studies* 44 (7), pp. 1377–1391.

Gurley, G. (2016). Black Lives Matter plunges into the affordable housing crisis. *American Prospect* September 2. http://prospect.org/article/black-lives-matter-plunges-affordable-housing-crisis Accessed November 29, 2016.

Ha, S-K. (2015) The endogenous dynamics of urban renewal and gentrification in Seoul. In Lees, L., Shin, H.B., López-Morales, E. *Global Gentrifications: Uneven Development and Displacement*. Bristol: Policy Press, pp. 165–180.

Haase, T. (2009) Divided City: The Changing Face of Dublin's Inner City. Dublin Inner City Partnership. http://trutzhaase.eu/wp/wp-content/uploads/R_2009_Divided-City.pdf Accessed February 9, 2016.

Halberstam, J. (2016) Who are "we" after Orlando? *Bully Bloggers*. June 22. https://bullybloggers.wordpress.com/2016/06/22/who-are-we-after-orlando-by-jack-halberstam/ Accessed August 31, 2016.

Hamilton, T. and Curran, W. (2013). From "five angry women" to "kick ass community": gentrification and environmental activism in Greenpoint, Brooklyn. *Urban Studies* 50(8), pp. 1557–1574.

Hamnett, C. and Butler, T. (2011) 'Geography matters': the role distance plays in reproducing educational inequality in East London. *Transactions of the Institute of British Geographers* 36, pp. 479–500, doi:10.1111/j.1475–5661.2011.00444.x

Hanhardt, C.B. (2013) *Safe Space: Gay Neighborhood History and the Politics of Violence*. Durham and London: Duke University Press.

Hankins, K. (2007) The final frontier: charter schools as the new community institutions of gentrification . *Urban Geography* 28(2), pp. 113–128.

Hannah-Jones, N. (2015) Gentrification doesn't fix inner city schools. *Grist*. February 27. http://grist.org/cities/gentrification-doesnt-fix-inner-city-schools/ Accessed July 19, 2016.

Haraway, D. (2013) Multispecies cosmopolitics: staying with the trouble. Institute for Humanities Research Distinguished Lecture, Institute for Humanities Research,

Arizona State University. 22 March. https://vimeo.com/62081248 Accessed October 14, 2016.

Harrison, T. (2012) Making the TTC and downtown hospitable to families. *Spacing Toronto*. July 26. http://spacing.ca/toronto/2012/07/26/making-the-ttc-and-down town-toronto-hospitable-to-families/ Accessed July 13, 2016.

Hayden, D. (2002) *Redesigning the American Dream: Gender, Housing, and Family Life*. New York and London: W.W. Norton and Co.

Heap, C. (2003) The city as a sexual laboratory: the queer heritage of the Chicago School. *Qualitative Sociology* 26(4), pp. 457–487.

Hearne, R. (2014) Actually existing neoliberalism: public-private partnerships in public space and infrastructure provision. In MacLaren, A. and Kelly, S., eds. *Neoliberal Urban Policy and the Transformation of the City: Reshaping Dublin*. New York: Palgrave Macmillan, pp. 157–173.

Hearne, R. and Redmond, D. (2014) The collapse of PPPs: Prospects for social housing regeneration after the crash. In MacLaren, A. and Kelly, S., eds. *Neoliberal Urban Policy and the Transformation of the City: Reshaping Dublin*. New York: Palgrave Macmillan, pp. 219–232.

Henig, J. R. (1984) Gentrification and Displacement of the Elderly: An Empirical Analysis, in Palen, J.J and London, B., eds. *Gentrification, Displacement and Neighborhood Revitalization*. Albany: State University of New York Press, pp. 173–185

Herbert, S. (2001) 'Hard charger' or 'station queen'? Policing and the masculinist state. *Gender, Place and Culture* 8(1), pp. 55–71.

Holloway, S. and Pimlott-Wilson, H. (2016) New economy, neoliberal state and professionalised parenting: mothers' labour market engagement and state support for social reproduction in class-differentiated Britain. *Transactions of the Institute of British Geographers*, doi:10.1111/tran.12130

Hubbard, P. (2004) Revenge and injustice in the neoliberal city: uncovering masculinist agendas. *Antipode*. 36(4), pp. 665–686.

Huse, T. (2014) *Everyday Life in the Gentrifying City*. Farnham: Ashgate.

Institute on Aging. (n.d.) http://www.ioaging.org/aging-in-america Accessed January 28, 2016.

Isoke, Z. (2014) Can't I be seen? Can't I be heard? Black women queering politics in Newark. *Gender, Place & Culture* 21(3), pp. 353–369.

Janoschka, M., Sequera, J. and Salinas, L. (2014) Gentrification in Spain and Latin America: a critical dialogue. *International Journal of Urban and Regional Research* 38(4), pp. 1234–1265.

Jezer-Morton, K. (2015) Are you a helicopter parent? Blame gentrification. *Next City*. August 17. https://nextcity.org/features/view/helicopter-parents-cities-gentrification Accessed July 11, 2016.

Joravsky, B. (2015) It's time Mayor Rahm treated teachers like they were police officer or firefighters. *Chicago Reader*. June 29. http://www.chicagoreader.com/Bleader/archives/2015/06/29/its-time-mayor-rahm-treated-teachers-like-they-were-police-offi cers-or-firefighters Accessed July 29, 2016.

Jupp, E. (2014) Women, communities, neighbourhoods: approaching gender and feminism within UK urban policy. *Antipode*. 46(5), pp.1304–1322.

Karsten, L. (2014) From Yuppies to YUPPs: family gentrifiers consuming spaces and re-inventing cities. *Tijdschrift voor Economische en Social Geografie* 105(2), pp. 175–188.

Katz, C. (2001) Vagabond capitalism and the necessity of social reproduction. *Antipode* 33, pp. 709–728, doi:10.1111/1467-8330.00207

Katz, C. (2008) Cultural geographies lecture: childhood as spectacle: relays of anxiety and the reconfiguration of the child. *Cultural Geographies* 15, pp. 5–17.

Keels, M., Burdick-Will, J. and Keene, S. (2013), The effects of gentrification on neighborhood public schools. *City & Community*, 12, pp. 238–259, doi:10.1111/cico.12027

Kelleher, M. (2015) Chicago leads way on charter school unions. *Catalyst Chicago.* November 6. http://catalyst-chicago.org/2015/11/chicago-lead-way-on-charter-school-unions/ Accessed July 28, 2016.

Kern, L. (2007) Reshaping the boundaries of public and private life: Gender, condominium development, and the neoliberalization of urban living. *Urban Geography* 28(7), pp. 657–681.

Kern, L. (2010a) *Sex and the Revitalized City: Gender, Condominium Development, and Urban Citizenship.* Vancouver: UBC Press.

Kern, L. (2010b) Selling the 'scary city': gendering freedom, fear and condominium development in the neoliberal city. *Social & Cultural Geography* 11 (3), pp. 209–230.

Kern, L. (2013) All aboard? Women working the spaces of gentrification in Toronto's Junction. *s* 20(4), pp. 510–527.

Kern, L. (2015) From toxic wreck to crunchy chic: environmental gentrification through the body. *Environment and Planning D: Society and Space* 33, pp. 67–83.

Kern, L. (2016) Rhythms of gentrification: eventfulness and slow violence in a happening neighbourhood. *Cultural Geographies* 23, pp. 441–457, doi:10.1177/1474474015591489

Kern, L. and Mullings, B. (2013) Urban neoliberalism, urban insecurity and urban violence: exploring the gender dimensions. In Peake, L. and Rieker, M., eds. *Rethinking Feminist Interventions into the Urban.* London and New York: Routledge, pp. 23–40.

Kim, K. (2006) Housing redevelopment and neighborhood change as a gentrification process in Seoul, South Korea: a case study of the Wolgok-4 Dong Redevelopment District. Electronic Theses, Treatises and Dissertations. Paper 3046.

Kinney, J. (2016) 10 Metros where rising rent is a big threat to America's seniors. *Next City.* February 1. https://nextcity.org/daily/entry/us-more-seniors-struggling-with-rent-affordable-housing Accessed February 2, 2016.

Klodawsky, F. and Spector, F. (1988) New families, new housing needs, new urban environments: the case of single-parent families. In Andrew, C. & Milroy, B. M., eds. *Life spaces: Gender, household, and employment.* Vancouver: University of British Columbia, pp. 141–158.

Knopp, L. (1995) Sexuality and urban space: a framework for analysis. In Bell, D. and Valentine, G., eds. *Mapping Desire: Geographies of Sexualities.* London and New York: Routledge, pp. 149–161.

Koenig, R. (2016) The landlord is a friend. *New York Times.* April 29. http://www.nytimes.com/2016/05/01/realestate/when-the-landlord-is-a-friend.html?_r=0 Accessed May 24, 2016.

Kotlowitz, A. (2015) Getting schooled. *New York Times Book Review.* August 23, pp. 9.

Kyung, S. (2011) 'State-facilitated gentrification' in Seoul, South Korea: for whom, by whom and with what result. Paper presented at the International RC21 Conference. Amsterdam. July 7–9. http://www.rc21.org/conferences/amsterdam2011/edocs/Session%202/2-1-Kyung.pdf. Accessed November 16, 2015.

Lang, N. (2012) When in Boystown: don't create racism, create change. *Huffington Post*. April 27. http://www.huffingtonpost.com/nico-lang/when-in-boystown_b_1457969.html Accessed August 24, 2016.

Lauria, M. and Knopp, L. (1985) Toward an analysis of the role of gay communities in the urban renaissance. *Urban Geography*, 6(2), pp. 152–169.

Lauster, N. and Easterbrook, A. (2011) No room for new families? A field experiment measuring rental discrimination against same-sex couples and single parents. *Social Problems* 58(3), pp. 389–409.

Lees, L. (2003) Super-gentrification: the case of Brooklyn Heights, New York City. *Urban Studies* 40(12), pp. 2487–2509.

Lees, L. (2016) Gentrification, race, and ethnicity: towards a global research agenda? *City & Community* 15(3), doi:10.1111/cico.12185

Lees, L., Slater, T., and Wyly, E. (2008) *Gentrification*. New York and London: Routledge.

Lees, L., Shin, H.B., López-Morales, E. (2015) *Global Gentrifications: Uneven Development and Displacement*. Bristol: Policy Press.

Leland, J. (2015a) 'The oldest old': a group portrait. *New York Times*. June 7. http://www.nytimes.com/2015/06/07/nyregion/a-group-portrait-of-new-yorks-oldest-old.html Accessed January 28, 2016.

Leland, J. (2015b) As lives lengthen, the costs mount. *New York Times*. November 15. http://www.nytimes.com/2015/11/15/nyregion/as-lives-lengthen-costs-mount.html Accessed January 28, 2016.

Lepore, J. (2016) Baby Doe: a political history of tragedy. *New Yorker* February 1: 46–57.

Levanon, A., England, P. and Allison, P. (2009) Occupational feminization and ppay: assessing causal dynamics using 1950–2000 U.S. Census data. *Social Forces* 88(2), pp. 865–891.

Ley, D. (1986) Alternative explanations for inner-city gentrification: a Canadian assessment. *Annals of the Association of American Geographers* 76(4), pp. 521–535.

Ley, D. (1987) The rent gap revisited. *Annals of the Association of American Geographers* 77(3), pp. 465–468.

Ley, D. (1996) *The New Middle Class and the Remaking of the Central City*. Oxford: Oxford University Press.

Ley, D. and Teo, S.Y. (2014) Gentrification in Hong Kong? Epistemology vs. ontology. *International Journal of Urban and Regional Research* 38(4), pp. 1286–1303.

Liner, E. (2016) A dollar short: what's holding women back from equal pay? *Third Way*. March 18. http://www.thirdway.org/report/a-dollar-short-whats-holding-women-back-from-equal-pay Accessed May 23, 2016.

Lipman, P. (2002) Making the global city, making inequality: the political economy and cultural politics of Chicago school policy. *American Educational Research Journal* 39(2), pp. 379–419.

Lipman, P. (2009) The cultural politics of mixed-income schools and housing: a racialized discourse of displacement, exclusion, and control. *Anthropology and Education Quarterly* 40(3), pp. 215–236.

Lipman, P. (2012) Mixed-income schools and housing policy in Chicago: A critical examination of the gentrification/ education/ 'racial' exclusion nexus. In Bridge, G., Butler, T. and Lees, L., eds. *Mixed Communities: Gentrification by Stealth?* Bristol: Policy Press, pp. 95–113.

Lipman, P. and Hursh, D. (2007) Renaissance 2010: the reassertion of ruling-class power through neoliberal policies in Chicago. *Policy Futures in Education* 5(2), pp. 160–178.

Loukaitou-Sideris, A. and Fink, C. (2009) Addressing women's fear of victimization in transportation settings: a survey of U.S. transit agencies. *Urban Affairs Review* 44(4), pp. 554–587.

Lowrey, A. (2016) Where did all the government jobs go? *New York Times Magazine.* May 1, pp. 64–67.

Lulay, S. (2016) Police checked out shots fired 1 hour before Aaren O'Connor was found shot. *DNAinfo Chicago.* February 10. https://www.dnainfo.com/chicago/20160210/pilsen/police-want-help-solving-murder-woman-shot-her-car-pilsen. Accessed February 23, 2016.

Lulay, S. and Nitkin, A. (2016) "Aaren O'Connor's killers" scrawled on Pilsen House. *DNAinfo Chicago,* February 15. https://www.dnainfo.com/chicago/20160215/pilsen/aaren-oconnors-killers-live-here-graffitti-on-pilsen-house-claims Accessed February 23, 2016.

Lyons, M. (1996) Employment, feminization, and gentrification in London, 1981–1993. *Environment and Planning A* 28, pp. 341–356.

Mackenzie, S. (1988) Building women, building cities: toward gender sensitive theory in the environmental disciplines. In Andrew, C. and Milroy, B.M., eds. *Life Spaces: Gender, Household, Employment.* Vancouver: University of British Columbia Press, pp. 13–64.

Mackenzie, S. and Rose, D. (1983) Industrial change, the domestic economy and home life. In Anderson, J., Duncan, S., and Hudson, R., eds. *Redundant Spaces in Cities and regions? Studies in Industrial Decline and Social Change.* London: Academic Press, pp. 155–200.

MacLaren, A. and Kelly, S., eds. (2014) *Neoliberal Urban Policy and the Transformation of the City: Reshaping Dublin.* New York: Palgrave Macmillan.

Malik, S., Barr, C., Holpuch, A. (2016) US millennials feel more working class than any other generation . *Guardian.* March 15. https://www.theguardian.com/world/2016/mar/15/us-millennials-feel-more-working-class-than-any-other-generation Accessed July 11, 2016.

Mallick, H. (2013) Strollers on the TTC: how else to take kiddies to the park? *The Star.* January 25. https://www.thestar.com/news/gta/2013/01/25/strollers_on_the_ttc_how_else_to_take_the_kiddies_to_the_park.html Accessed July 13, 2016.

Manalansan, M. F. (2015), Queer worldings: the messy art of being global in Manila and New York. *Antipode,* 47(3), pp. 566–579, doi:10.1111/anti.12061

Marcuse, P. (1997) The enclave, the citadel, and the ghetto: what has changed in the Post-Fordist U.S. city. *Urban Affairs Review* 33(2), pp. 228–264.

Margolis, E. (2015) Closing time: the loss of iconic gay venues is a nasty side-effect of London's sanitization. *New Statesman.* March 11. http://www.newstatesman.com/politics/2015/03/closing-time-loss-iconic-gay-venues-nasty-side-effect-londons-sanitisation Accessed August 27, 2016.

Markusen, A. (1981) City spatial structure, women's household work, and national urban policy. In Stimpson, C., Dixler, E., Nelson, M.J. and Yatrakis, K.B., eds. *Women and the American City.* Chicago: University of Chicago Press, pp. 20–41.

Marshall, A. (2014) The majority of urban parents in the U.S. say they struggle to meet household expenses. *CityLab.* September 3. http://www.citylab.com/work/2014/09/the-majority-of-urban-parents-in-the-us-say-they-struggle-to-meet-household-expenses/379550/ Accessed July 13, 2016.

Martin, L. (2008) Boredom, drugs, and schools: protecting children in gentrifying communities. *City & Community* 7(4), pp. 331–346.

Martinez, A. (2015) Queer cosmopolis: the evolution of Jackson Heights. In Doan, P., ed. *Planning and LGBTQ Communities: The Need for Inclusive Queer Spaces.* New York and London: Routledge, pp. 167–180.

McDowell, L. (1983) Towards an understanding of the gender division of urban space. *Environment and Planning D: Society and Space* 1, pp. 59–72.

McDowell, L. (2003) *Redundant Masculinities? Employment Change and White Working Class Youth.* Oxford: Blackwell.

McDowell, L. (2014) The lives of others: body work, the production of difference, and labor geographies. *Economic Geography* 91(1), pp. 1–23.

McGhee, J. (2016) Saying good-bye to Girlstown: Andersonville's lesbian population shrinks. *DNAInfo Chicago.* August 22. https://www.dnainfo.com/chicago/20160822/a ndersonville/saying-goodbye-girlstown-andersonvilles-lesbian-population-shrinks Accessed August 25, 2016.

McLean, H., Rankin, K., Kamizaki, K. (2015) Inner-suburban neighborhoods, activist research, and the social space of the commercial street. *ACME: An International E-Journal for Critical Geographies* 14(4), pp. 1283–1308.

McNeil, J. (2015) Why do I have to call this app 'Julie'? *New York Times.* December 20, pp. SR 4.

Melt, H. (2015) Why Black queer activists engaged in civil disobedience at Chicago Gay Pride Parade. *In These Times.* July 9. http://inthesetimes.com/article/18177/why-bla ck-queer-activists-engaged-in-civil-disobedience-at-chicago-gay-prid Accessed August 29, 2016.

Meyer, B. and Mittag, N. (2015) Using linked survey and administrative data to better measure income: implications for poverty, program effectiveness and holes in the safety net. National Bureau of Economic Research Working Paper No. 21676. http://www.nber.org/papers/w21676.pdf Accessed January 21, 2016.

Miller, C.C. (2015) Men do more at home, but not as much as they think. *New York Times.* November 12. http://www.nytimes.com/2015/11/12/upshot/men-do-more-a t-home-but-not-as-much-as-they-think-they-do.html?_r=0. Accessed November 16, 2015.

Miller, C.C. (2016) As women take over a male-dominated field, the pay drops. *New York Times.* March 18. http://www.nytimes.com/2016/03/20/upshot/as-women-take -over-a-male-dominated-field-the-pay-drops.html?_r=0 Accessed May 23, 2016.

Mills, C. (1988) Life on the upslope: the postmodern landscape of gentrification. *Environment and Planning D: Society and Space* 6, pp. 169–189.

Miranne, K.B. (2000) Women 'embounded': Intersections of welfare reform and public housing. In Miranne, K.B. and Young, A.H., eds. *Gendering the City: Women, Boundaries and Visions of Urban Life.* Lanham: Rowman and Littlefield.

Mock, B. (2015a) We are living in the era of job gentrification. *CityLab* September 3. http://www.citylab.com/work/2015/09/the-era-of-job-gentrification/403492/ Accessed September 8, 2015.

Mock, B. (2015b) The scourge of sexual harassment in public housing . *CityLab.* November 16. http://www.citylab.com/housing/2015/11/the-scourge-of-sexual-hara ssment-in-public-housing/415908/?utm_source=nl__link2_111715. Accessed March 3, 2016.

Monroe, I. (2013) Has gentrification of New York's Harlem neighborhood stirred up prejudice against LGBTQ residents? *Huffington Post.* September 7. http://www. huffingtonpost.com/irene-monroe/harlem-gentrification_b_3877218.html Accessed August 25, 2016.

Moser, C. (2016) Making gender equality central to the New Urban Agenda. *Next City*. October 6. https://nextcity.org/daily/entry/gender-equality-new-urban-agenda Accessed October 20, 2016.

Mountz, A. and Curran, W. (2009) Policing in drag: Giuliani goes global with the illusion of control. *Geoforum* 40, pp. 1033–1040.

Muñiz, V. (1998) *Resisting Gentrification and Displacement: Voices of the Puerto Rican Women of the Barrio*. New York: Garland.

Muñoz, J.E. (2009) *Cruising Utopia: The Then and There of Queer Futurity*. New York and London: New York University Press.

Namaste, K. (1996) Genderbashing: sexuality, gender, and the regulation of public space. *Environment and Planning D: Society and Space* 14, pp. 221–240.

Nash, C. J. and Gorman-Murray, A. (2014) LGBT neighbourhoods and 'new mobilities': towards understanding transformations in sexual and gendered urban landscapes. *International Journal of Urban and Regional Research* 38: 756–772, doi:10.1111/1468-2427.12104Nast, H. (2002) Queer patriarchies, queer racisms, international. *Antipode* 34(5): 874–909.

Nast, H. (2002) Queer patriarchies, queer racisms, international. *Antipode* 34(5): 874–909.

Nast, H. (2016) Into the arms of dolls: Japan's declining fertility rates, the 1990s financial crisis and the (maternal) comforts of the posthuman. *Social and Cultural Geography*, doi:10.1080/14649365.2016.1228112

National Low Income Housing Coalition. (2012) Who lives in federally assisted public housing? *Housing Spotlight* 2(2). November. http://nlihc.org/sites/default/files/Hou singSpotlight2-2.pdf Accessed January 19, 2016.

Newman, K. and Wly, E. (2006) The right to stay put, revisited: gentrification and resistance to displacement in New York City. *Urban Studies* 43, 1: 23–57.

Noterman, E. (2016) Beyond tragedy: differential commoning in a manufactured housing cooperative. *Antipode*, 48, pp. 433–452, doi:10.1111/anti.12182

Nyden, P., Edlynn, E., and Davis, J. (2006) The differential impact of gentrification on communities in Chicago. Loyola University Chicago Center for Urban Research and Learning for the City of Chicago Commission on Human Relations. January. http://www1.luc.edu/media/lucedu/curl/pdfs/HRC_Report.pdf Accessed January 28, 2016.

Older, D.J. (2014) Gentrification's insidious violence: the truth about American cities. *Salon*. April 8. http://www.salon.com/2014/04/08/gentrifications_insidious_violence_the_truth_about_american_cities/ Accessed October 21, 2016.

Papadopoulos, A.G. (2017) Becoming "Boystown" in neoliberal Chicago: a critical urban morphology of the North Halsted-Broadway Corridor. In Bennett, L., Garner, R. and Hague, E., eds. *Neoliberal Chicago*. Urbana-Champaign: University of Illinois Press, pp. 161–190.

Papayanis, M. (2000) Sex and the revanchist city: zoning out pornography in New York. *Environment and Planning D: Society and Space* 18, pp.341–354

Park, Y.J., Kim, K. and Jeon, H.R. (2010) Working women's preferences for a residential location choice: a case study in Seoul, Korea. *Housing and Society* 37(2), pp. 143–157.

Parker, B. (2008) Beyond the class act: Gender and race in the "Creative City' discourse. In DeSena, J., ed. *Gender in an Urban World*. Bingley: Emerald Publishing Group, pp. 201–232.

Parker, B. (2016a). Feminist forays in the city: imbalance and intervention in urban research methods. *Antipode*, doi:10.1111/anti.12241

Parker, B. (2016b) The feminist geographer as killjoy: excavating gendered urban power relations, *The Professional Geographer*, doi:10.1080/00330124.2016.1208513

Patch, J. (2008) "Ladies and gentrification": new stores, residents, and relationships in neighborhood change. *Gender in an Urban World, Research in Urban Sociology*, 9, pp.103–126.

Pattillo, M. (2007) *Black on the Block: The Politics of Race and Class in the City – A New Perspective of Middle Class*. Chicago: University of Chicago Press.

Paton, K. (2014) *Gentrification: A Working-Class Perspective*. Farnham: Ashgate.

Peake, L. and Rieker, M., eds. (2013) *Rethinking Feminist Interventions into the Urban*. London and New York: Routledge.

Pearsall, H., 2010. From brown to green? Assessing social vulnerability to environmental gentrification in New York city. *Environment and Planning C*, 28(5), pp. 872–886.

Peck, J. (2005) Struggling with the creative class. *International Journal of Urban and Regional Research* 29(4), pp. 740–770.

Peck, J. (2012) Austerity urbanism. *City* 16(6), pp. 626–655.

Petrovic, A. (2008) The elderly facing gentrification: neglect, invisibility, entrapment, and loss. *Elder Law Journal* 15, pp. 533–579.

Podmore, J. (2006) Gone underground? *Social and Cultural Geography* 7, pp. 595–625.

Podmore, J. (2013) Critical commentary: sexualities landscapes beyond homonormativity. *Geoforum* 49, pp. 263–267.

Portacolone, E. and Halpern, J. (2014) "Move or suffer": is age segregation the new norm for older Americans living alone? *Journal of Applied Gerontology*, doi:10.1177/0733464814538118

Pratt, G. (2004) *Working Feminism*. Philadelphia: Temple University Press.

Preville, P. (2014) Stuck in Condoland. *Toronto Life*. June 11. http://torontolife.com/city/stuck-in-condoland/ Accessed January 12, 2016.

Pruitt-Igoe Myth. (2011) Chad Friedrichs, dir.

Prynn, J. (2016) Super-rich foreigners 'forcing the old money elite out of London's prime postcodes.' *Evening Standard*. August 31. http://www.standard.co.uk/news/london/superrich-foreigners-forcing-the-old-money-elite-from-londons-prime-postco des-a3332961.html Accessed November 8, 2016.

Quart, A. (2013) Crushed by the costs of child care. *New York Times*. August 18: SR 4.

Quastel, N., (2009). Political ecologies of gentrification. *Urban Geography*, 30(7), pp. 694–725.

Rahder, B. and McLean, H. (2013) Other ways of knowing your place: immigrant women's experience of public space in Toronto. *Canadian Journal of Urban Research* 22(1), pp. 145–166.

Rankin, K.N. and McLean, H. (2015) Governing the commercial streets of the city: new terrains of disinvestment and gentrification in Toronto's inner suburbs. *Antipode* 47(1), pp. 216–239, doi:10.111/anti.120996

Richie, B.E. (2012) *Arrested Justice: Black Women, Violence, and America's Prison Nation*. New York: New York University Press.

Rose, D. (1984) Rethinking gentrification: Beyond the uneven development of Marxist urban theory. *Environment and Planning D: Society and Space* 1, pp. 47–74.

Rose, D. (1989) Feminist perspective of employment restructuring and gentrification: the case of Montréal. In Wolch, J. and Dear, M., eds. *The Power of Geography*. Boston: Unwin Hyman, pp. 118–138.

Rose, D. (2010) Refractions and recombinations of the 'economic' and the 'social': a personalized reflection on challenges by-and to- feminist urban geographies. *Canadian Geographer* 54(4), pp. 391–409.

Ross, S. (2014) New faces, old parties – political patronage goes on. *Irish Independent*. June 1. http://www.independent.ie/business/irish/new-faces-old-parties-political-pa tronage-goes-on-30319085.html Accessed July 8, 2016.

Rothenberg, T. (1995) 'And she told two friends': lesbians creating urban social space. In Bell, D. and Valentine, G., eds. *Mapping Desire: Geographies of Sexualities*. London and New York: Routledge, pp. 165–181.

Ruddick, S. (1992) Thinking about fathers. In Thorne, B. and Yalom, M., eds. *Rethinking the Family: Some Feminist Questions*. Boston: Northeastern University Press. pp 176–190.

Safransky, S. (2014) Greening the urban frontier: race, property, and resettlement in Detroit. *Geoforum* 56(4), pp. 237–248.

Sanchez, C. (2006) Building power. *Chicago Reporter* 35, pp. 6–15.

Sandberg, L.A. (2014) Environmental gentrification in a post-industrial landscape: the case of the Limhamn quarry, Malmö, Sweden. *Local Environment* 19(10), pp. 1068–1085.

Satow, J. (2015) A trophy tower? Nah. *New York Times*. November 15.

Schulman, S. (2012) *The Gentrification of the Mind: Witness to a Lost Imagination*. Berkeley: University of California Press.

Searcey, D. and Bradsher, K. (2015) The great sprawl. *New York Times*. November 29.

Shaw, K. (2012) Beware the Trojan horse: social mix constructions in Melbourne. In Bridge, G., Butler, T. and Lees, L. *Mixed Communities: Gentrification by Stealth?* Bristol: Policy Press, pp. 133–148.

Shaw, K. and Hagemans, I. (2015) 'Gentrification without displacement' and the consequent loss of place: the effects of class transition on low-income residents of secure housing in gentrifying areas. *International Journal of Urban and Regional Research*, 39: 323–341, doi:10.1111/1468–2427.12164

Shaw, W.S. (2007) *Cities of Whiteness*. Malden, MA: Blackwell.

Shin, H.B. (2009) Property-based redevelopment and gentrification: the case of Seoul, South Korea. *Geoforum* 40(5), pp. 906–917.

Shin, H.B. and Kim, S-H. (2015) The developmental state, speculative urbanization and the politics of displacement in gentrifying Seoul. *Urban Studies*, doi:10.1177/ 004209801 4565745

Siltanen, J., Klodawsky, F. and Andrew, C. (2015) 'This is how I want to live my life': an experiment in prefigurative feminist organizing for a more equitable and inclusive city. *Antipode* 47(1), pp. 260–279, doi:19.1111/anti.12092

Slater, T. (2006) The eviction of critical perspectives from gentrification research . *International Journal of Urban and Regional Research* 30(4), pp. 303–325.

Slaughter, A. (2015) A toxic work world. *New York Times*. September 20, pp. SR 1;6.

Smith, J. (2005) Housing, gender, and social policy. In Somerville, P. with Sprigings, N., eds. *Housing and Social Policy: Contemporary Themes and Critical Perspectives*. New York: Routledge, pp. 143–171.

Smith, J. (2006) The Chicago Housing Authority's plan for transformation. In Bennett, Larry, Smith, Janet L. and Wright, Patricia A., eds. *Where Are Poor People to Live? Transforming Public Housing Developments*. Armonk,NY: M.E. Sharpe, pp. 93–124.

Smith, J. (2015) Between a rock and a hard place: public housing policy. *Journal of Urban Affairs* 37(1), pp. 42–46.

Smith, N. (1979) Toward a theory of gentrification: a back to the city movement by capital, not people . *Journal of the American Planning Association* 45(4), pp. 538–548.

Smith, N. (1987) Gentrification and the rent gap. *Annals of the Association of American Geographers* 77(3), pp. 462–478.

Smith, N. (1996) *The New Urban Frontier: Gentrification and the Revanchist City.* London and New York: Routledge.

Smith, N. (2002). New globalism, new urbanism: gentrification as global urban strategy. *Antipode* 34(3): pp. 427–450.

Spain, D. (2001) *How Women Saved the City.* 2001. Minneapolis: University of Minnesota Press

Sosin, K. (2012) Center on Halsted offers youth services ... but not without controversy. *Windy City Times.* December 19. http://www.windycitymediagroup.com/lgbt/Center-on-Halsted-offers-youth-servicesbut-not-without-controversy/40875.html Accessed August 24, 2016.

Stamp, J. (1980) Towards supportive neighborhoods: women's role in changing the segregated city. In Wekerle, G., Peterson, R. and Morley, D., eds. *New Space for Women.* Boulder, CO: Westview Press, pp. 189–198.

Stand Up! Chicago. (2012) Fight for the future: how low wages are failing children in Chicago's schools. http://fightfor15chicago.org/wordpress/wp-content/uploads/2014/03/Education-Report-FINAL.pdf Accessed July 27, 2016.

Stewart-Winter, T. (2015) The price of gay marriage. *New York Times.* June 28, pp. SR1; 6.

Stewart-Winter, T. (2016) *Queer Clout: Chicago and the Rise of Gay Politics.* Philadelphia: University of Pennsylvania Press.

Strauss, V. (2015) Why hunger strikers are risking their health to save a Chicago public high school. *Washington Post.* August 29. https://www.washingtonpost.com/news/answer-sheet/wp/2015/08/29/why-hunger-strikers-are-risking-their-health-to-save-a-chicago-public-high-school/ Accessed July 22, 2016.

Stone, J. (2015) Why march for homes? Because the housing crisis goes far beyond us Focus E15 mums. *The Guardian.* January 31. http://www.theguardian.com/commentisfree/2015/jan/31/march-for-homes-focus-e15-mums-london-homelessness-priced-out-area Accessed February 9, 2016.

Sweet, E.L. (2016) Gender, violence, and the city of emotion. In Beebeejaun, Y., ed. *The Participatory City.* Berlin: Jovis, pp. 120–127.

Taylor, Y. (2013) 'That's not really my scene': working-class lesbians in (and out of) place. In Taylor, Y. and Addison, M., eds. *Queer Presences and Absences.* New York: Palgrave Macmillan, pp. 159–177.

Taylor, Y. and Addison, M., eds. (2013) *Queer Presences and Absences.* New York: Palgrave Macmillan.

Teece-Johnson, D. and Burton-Bradley, T. (2016) Inner Sydney's Aboriginal community fear they are being pushed out for 'white hipsters.' *NITV.* March 10. http://www.sbs.com.au/nitv/the-point-with-stan-grant/article/2016/03/09/inner-sydneys-aboriginal-community-fear-they-are-being-pushed-out-white-hipsters Accessed December 9, 2016.

Terrell, M. (2016) The battle to save San Francisco's queer spaces from gentrification. *Vice.* July 19. http://www.vice.com/read/how-to-save-san-franciscos-queer-bars-from-gentrification Accessed August 27, 2016.

Tharoor, I. 2016. Black Lives Matter is a global cause. *Washington Post.* July 12. https://www.washingtonpost.com/news/worldviews/wp/2016/07/12/black-lives-matter-is-a-global-cause/ Accessed November 29, 2016.

Tickell, A. and Peck, J. (1996) The return of the Manchester Men: men's words and men's deeds in the remaking of the local state. *Transactions of the Institute of British Geographers*, 21(4), pp. 595–616.

TransX Istanbul. (2014). Maria Binder, dir.

Tretter, E. (2013) Contesting sustainability: 'SMART growth' and the redevelopment of Austin's Eastside. *International Journal of Urban and Regional Research* 37(1), pp. 297–310.

Valentine, D. (2007) *Imagining Transgender: An Ethnography of a Category.* Durham and London: Duke University Press.

van den Berg, M. (2012) Femininity as a city marketing strategy: gender bending Rotterdam. *Urban Studies* 49(1), pp. 153–168.

van den Berg, M. (2013) City children and genderfied neighborhoods: the new generation of urban regeneration strategy. *International Journal of Urban and Regional Research* 37(2), pp. 523–536.

Vigdor, J. (2002) Does gentrification harm the poor? *Brookings-Wharton Papers on Urban Affairs*, pp. 133–173.

Visser, G. (2013) Challenging the gay ghetto in South Africa: time to move on? *Geoforum* 49, pp. 268–274.

Warde, A. (1991) Gentrification as consumption: issues of class and gender. *Environment and Planning D: Society and Space* 9, pp. 223–232.

Warner, M.E. and Prentice, S. (2012) Regional economic development and child care: toward social rights. *Journal of Urban Affairs* 35(2), pp. 195–217.

Warren, E. and Tyagi, A.W. (2003) *The Two-Income Trap: Why Middle-Class Parents are Going Broke.* New York: Basic Books.

Watt, P. (2016) A nomadic war machine in the metropolis, *City*, 20(2), pp. 297–320, doi:10.1080/13604813.2016.1153919

Weber, R. (2015) *From Boom to Bubble: How Finance Built the New Chicago.* Chicago and London: University of Chicago Press.

Wekerle, G. (1984) A woman's place is in the city. *Antipode* 16, pp. 11–19.

Wekerle, G. (2003) From eyes on the street to safe cities. *Places* 13, pp. 44–49

Wekerle, G. (2013) Interrogating gendered silences in urban policy. In Peake, L. and Rieker, M., eds. *Rethinking Feminist Interventions into the Urban.* London and New York: Routledge, pp. 142–158.

Whyte, W. (1988) *City: Rediscovering the Center.* New York: Anchor Books.

Wichterich, C. (2015) Contesting green growth, connecting care, commons and enough. In Harcourt, W., and Nelson, I.L., eds. *Practising Feminist Political Ecologies: Moving Beyond the 'Green Economy.'* London: Zed Books, pp. 67–100.

Williams, B. and Redmond, D. (2014) Ready money: residential over-development and its consequences. In MacLaren, A. and Kelly, S., eds. *Neoliberal Urban Policy and the Transformation of the City: Reshaping Dublin.* New York: Palgrave Macmillan, pp. 107–119.

Wilson, D. and Grammenos, D. (2005). Gentrification, discourse, and the body: Chicago's Humboldt Park. *Environment and Planning D: Society and Space* 23: 295–312.

Wilson, D., Wouters, J., and Grammenos, D. (2004) Successful protect-community discourse: spatiality and politics in Chicago's Pilsen neighborhood. *Environment and Planning A* 36, pp. 1173–1190.

Wolfe, M. (1992) Invisible women in invisible places: lesbians, lesbian bars, and the social production of people/environment relationships. *Architecture and Behavior* 8(2), pp. 137–158.

Woodstock Institute. (2015) Her longer road home: disparities in mortgage lending in the Chicago region. Woodstock Institute. June. http://www.woodstockinst.org/resea rch/her-longer-road-home-disparities-mortgage-lending-women-chicago-region. Accessed December 1, 2015.

Wright, M.W. (2004) From protests to politics: sex work, women's worth, and Ciudad Juarez Modernity. *Annals of the Association of American Geographers* 94(2), pp. 269–286.

Wright, M.W. (2014) Gentrification, assassination and forgetting in Mexico: a feminist Marxist tale. *Gender, Place and Culture* 21(1), pp. 1–16.

Wyly, E. and Hammel, D. (1999) Islands of decay in seas of renewal: housing policy and the resurgence of gentrification. *Housing Policy Debate* 10(4), pp. 711–771

Wyly, E., Newman, K., Schafran, A. and Lee, E. (2010) "Displacing New York." *Environment and Planning A* 42(11), pp. 2602–2623, doi:10.1068/a42519

Yerkes, M., Standing, K., Wattis, L., and Wain, S. (2010) The disconnection between policy practices and women's lived experiences: combining work and life in the UK and the Netherlands. *Community, Work & Family* 13(4), pp. 411–427.

Zehr, M.A. (2009) Nurturing 'school minds'. *Education Week*. October 2. http://www. edweek.org/ew/articles/2009/10/07/06uno_ep.h29.html Accessed July 19, 2016.

Zuk, M., Bierbaum, A.H., Chapple, K., Gorska, K., Loukaitou-Sideris, A., Ong, P., and Thomas, T. (2015) Gentrification, displacement and the role of public investment: a literature review. Federal Reserve Bank of San Francisco Community Development Invest Center Working Paper. http://www.frbsf.org/communi ty-development/files/wp2015-05.pdf

Zukin, S. (2016), Gentrification in three paradoxes. *City & Community*, 15, pp. 202–207, doi:10.1111/cico.12184

Index

Page numbers in *italics* denote figures.